MW00464687

The STAR *of*
BETHLEHEM

DEDICATED TO
DR MICHAEL ROBERT MOLNAR
(1945-2023)
WHO STARTED ME ON THIS JOURNEY.

Nihil Obstat: The Reverend Mark S. Stelzer, STD
  Diocese of Springfield, Massachusetts
  8th July 2023

Imprimatur: The Reverend Monsignor Gerald Ewing, VG
  Archdiocese of Southwark
  19th July 2023

The *Nihil Obstat* and *Imprimatur* are a declaration that a publication is considered to be free from doctrinal or moral error. It is not implied that those who have granted the *Nihil Obstat* and *Imprimatur* agree with the contents, opinions or statements expressed.

The STAR *of* BETHLEHEM

What Did the Magi See?

Fr Douglas McGonagle

All booklets are published
thanks to the generosity of the supporters
of the Catholic Truth Society

Acknowledgements

Some material in this work originally appeared in *The Catholic Mirror*, a magazine of the Diocese of Springfield, December 2013-January 2014 issue.

I am indebted to my good friend Gordon W. Tinkham Jr for graciously spending time reading this text and so helping make it intelligible. I wish to thank Michael M. Serafin for his assistance in running to ground a half-remembered C.S. Lewis quotation. Finally, I warmly thank the parishioners of Most Holy Redeemer Parish for their forbearance at having a part-time pastor while I worked on this project.

Rev Douglas McGonagle
Hadley, Massachusetts

All rights reserved. First published 2023 by The Incorporated Catholic Truth Society, 42-46 Harleyford Road, London SE11 5AY. Tel: 020 7640 0042. © 2023 The Incorporated Catholic Truth Society.
www.ctsbooks.org

ISBN 978 1 78469 764 8

CONTENTS

Introduction

And the rafters of toil still are gilded
With the dawn of the star of the heart,
And the wise men draw near in the twilight,
Who are weary of learning and art,
And the face of the tyrant is darkened,
His spirit is torn,
For a new king is enthroned; yea, the sternest,
A child is born.

G.K. Chesterton, *The Nativity*

Years ago, when people found out that I was an astronomer as well as a Roman Catholic priest, they would often question me about the Star of Bethlehem.

'What was it?" they would ask.

"I don't have the foggiest," I would reply.

A bright star shining in the night sky, its rays of light streaming down onto a stable in Bethlehem, is a classic Christmas tableau, and yet the star only appears in one brief and puzzling account in the Gospel of Matthew. As Pope Benedict XVI notes in his book *Jesus of Nazareth:*

The Infancy Narratives, "Hardly any biblical narrative has so caught the imagination or stimulated so much research and reflection as the account of the "Magi' from the "land of the sunrise", which the evangelist Matthew adds directly after the story of Jesus's birth."[1]

So, let us start our research by reading the short biblical narrative from the Gospel of Matthew 2:1-12 where we learn all there is to know about the star and the Magi:

Now when Jesus was born in Bethlehem of Judea in the days of Herod the king, behold, Wise Men from the East came to Jerusalem, saying, "Where is he who has been born king of the Jews? For we have seen his star in the East, and have come to worship him." When Herod the king heard this, he was troubled, and all Jerusalem with him; and assembling all the chief priests and scribes of the people, he inquired of them where the Christ was to be born. They told him, "In Bethlehem of Judea; for so it is written by the prophet:

'And you, O Bethlehem, in the land of Judah,
are by no means least among the rulers of Judah;
for from you shall come a ruler
who will govern my people Israel.'"

Then Herod summoned the Wise Men secretly and ascertained from them what time the star appeared;

[1] Pope Benedict XVI, *Jesus of Nazareth: The Infancy Narratives* (New York: Image, 2012) p. 89.

and he sent them to Bethlehem, saying, "Go and search diligently for the child, and when you have found him bring me word, that I too may come and worship him." When they had heard the king they went their way; and behold, the star which they had seen in the East went before them, till it came to rest over the place where the child was. When they saw the star, they rejoiced exceedingly with great joy; and going into the house they saw the child with Mary his mother, and they fell down and worshiped him. Then, opening their treasures, they offered him gifts, gold and frankincense and myrrh. And being warned in a dream not to return to Herod, they departed to their own country by another way.

And that is all we ever hear about a star. As popular and thought-provoking an image as it is, it is all the first-hand information we have about the star.

And a strange star it is.

First of all, the star is conspicuous enough to catch the attention of the Magi – so conspicuous, in fact, that they packed up and headed off to Jerusalem to find the newborn king of the Jews. Yet the star is seemingly invisible to Herod, the chief priest and the scribes, and all of Jerusalem. Apparently, the arrival of the Magi is the first they have heard of any star.

The star also has some peculiar properties: it rises in the east, but then proceeds before the Magi as they travel west, before stopping over the place where the child was. Not many stars in my experience can pull off those tricks. But wait a minute. If the star hovered over the place where Jesus was, why, then, did the Magi go first to Jerusalem to ask where the new King was to be born?

It is all very puzzling, and many people, over the centuries, have tried to solve the riddle of this most peculiar star.

Now, we moderns are biased towards wondering what the star was physically. We wonder whether it could it have been a comet, or a close clustering of bright objects in the night sky such as the Moon, Jupiter, and Venus. Or might it even have been a supernova? All these modern speculations reveal our bias towards associating important events with visually spectacular displays.

Personally, I had no strong preference for one physical explanation for the star over another. How the star manifested itself was always secondary. To me it always seemed that the star's importance derived from the role it played astrologically. After all, the Magi, the fellows who show up in Jerusalem asking for the whereabouts of the new-born king of the Jews, were astrologers. Since the Wise Men were astrologers, it always seemed logical to me that the star's significance lay in it being an astrological omen of a regal birth. It might still have been an impressive visual display, perhaps even one worthy of a big budget sci-fi movie, but even so the display would remain secondary to what the star signified: the birth of a king.

Furthermore, for me as a Christian, the star's physical properties were always secondary to what I believed to be the important Gospel message Matthew was trying to convey – namely, that even Gentiles, represented by pagan astrologers from the east, recognised that Jesus was God. Matthew writes about the Magi finally arriving in Bethlehem, describing how, on entering the house, "they saw the child with Mary his mother and fell down and they worshipped" (*Mt* 2:11). Today, we carelessly throw around words like "worship"; however, for Matthew, "worshipped" (*προσεκύνησαν*)[2] is used to refer to the divine.

Matthew will use the word "worshipped" again at the close of his Gospel. After the resurrection, the disciples go to Galilee, to the mountain to which Jesus had directed them, and "when they saw Him, they worshipped (*προσεκύνησαν*) Him; but some doubted." (*Mt* 28:17) Matthew, who is trying to convince his readers that Jesus of Nazareth is truly the Messiah, bookends his Gospel story with people giving worship to Jesus. Curiously, the Gentile Magi appear to do so spontaneously and without question, whereas his Jewish disciples – his own friends – gave Jesus worship, yet still doubted his divinity.

Again, for me, the physical manifestation of the star was always secondary to the notion that the very "powers of the heavens" (*Mt* 24:29) took note of the Incarnation and

---

[2] Nestle, E. and McReynolds, P.R., *Nestle-Aland 26th Edition Greek New Testament with McReynolds English Interlinear* (Logos Research Systems, Inc., 1997) (*Mt* 2:11).

arranged themselves in such a way as to draw the Gentiles, represented by the Magi, into the presence of the one true God. That is, it was secondary to me, until I read the book *The Star of Bethlehem: The Legacy of the Magi*, by Dr Michael Molnar, then at Rutgers University.[3]

Dr Molnar's interest in the star of Bethlehem began quite by accident. He was an amateur collector of ancient coins, and in the spring of 1991 he was trying to understand the symbolism on several Roman provincial coins minted in the city of Antioch around the year AD 6, a time coincident with Syria being governed by a man named Quirinius. On one side of the coin is an image of Zeus: no surprise there. However, on the other side there is the strange image of a leaping ram looking backwards at a star. Molnar noted that the scene is intriguing if one recalls the Gospel of Luke, which tells us that Jesus was born when Quirinius was governor of Syria (Lk 2:2). Could this coin have some connection with the star that the Magi followed as reported in Matthew's Gospel?

On reading this, I too became intrigued. Whereas previously the physical nature of the star was only a mere curiosity to me, now that there might be hard physical evidence for the star's historicity I became very interested in the star's physical phenomena.

---

[3] Molnar, Michael R., *The Star of Bethlehem: The Legacy of the Magi* (New Brunswick, NJ: Rutgers University Press, 1999).

By the time I finished reading his book, I was convinced that Dr Molnar was certainly on to something. Whether or not the coins are truly linked to the star we may never know with metaphysical certitude. The English writer and Christian apologist C.S. Lewis reminds us that all historical facts are, in fact, taken on authority – that is, we believe them true because we find that the people who tell us that such and such happened at this or that time in one place or another are themselves trustworthy. We find the preponderance of evidence to be consistent with the historical fact being true.

Based on the evidence and arguments presented in Dr Molnar's book, I am quite willing to believe that the coins could be linked to the Star of Bethlehem and are a "legacy of the Magi". I say "could be", because, as a trained scientist like Dr Molnar, I know that it only takes one inconsistent datum to destroy the most cherished of theories. Scientific theories, whether astronomical or historical, are always tentative. As a scientist, I know it is dangerous to get too attached to a theory. As one of my undergraduate professors, Dr Thomas Arny, liked to tell us, "Science is doomed to success." If a theory is found wanting, he explained, we simply throw it out and replace it with a new theory that is consistent with all the known data.

Another thing my professors taught me was to never make a declarative statement. Every statement needs an escape clause, such as "the data would seem to show…", "the evidence appears to be consistent with…" or "with continued funding we intend to demonstrate that…". My instructors were so effective in drilling this into me that, to this very day, I become wary whenever a declarative statement is made, especially when the statement purports to be a declaration of scientific fact. Science is never settled; it is always provisional (and, yes, this is a declarative as well as a possibly recursive statement).

Provisional or not, I felt that Dr Molnar was on to something. I decided to write a public talk – a book report, really – that I could deliver to groups and so get the word out about the possible discovery of material evidence for the Star of Bethlehem. It was the autumn of 2004, and I was a parochial vicar in a suburban parish. My first talks, complete with PowerPoint eye candy, were given to my parishioners. Soon I was being invited to parishes in my Diocese, usually during Advent and Christmastide, to give what was becoming known as the "Star Talk". I have given the Star Talk in many venues, including a science museum and a Benedictine abbey – both in the same week. The Talk was so well received at one parish that they decided to memorialise it on the ceiling of the sanctuary that they were restoring at the time. I sent them the star chart from the Star Talk, and an artist painted it on the ceiling, where it can still be seen today.

I must warn the reader that we PhDs are notorious for thinking that just because we are knowledgeable in one topic, we have licence to opine on any topic. Though

I am not a trained historian, I am a fan of history who enjoys speculating about historical events and trends. As a diocesan priest who has taken a few courses in philosophy, I readily admit that speaking about philosophy does give me some pause. Several close friends of mine are *real* philosophers before whom I would rather not embarrass myself; nevertheless, under extreme social pressure, I have been known to opine on things philosophical and so risk their chagrin.

I take the time to relate all this so that readers may be aware that, in the discussions that follow, I am relying heavily on the original work done by Dr Molnar as described in his book, *The Star of Bethlehem: The Legacy of the Magi*. For students of the Classics among you, I invite you to think of Dr Molnar as my Virgil as in Dante's *Divine Comedy*.

That said, I do believe that I have some original thoughts and insights that may advance the topic. Years of people probing the depths of my ignorance with their questions, at the conclusion of my Star Talks, has motivated me to learn more about the history, philosophy and astronomy, as well as the astrology, that underlies the topic. Often, their questions have uncovered new avenues of inquiry to be explored. And so...

Intrigued by a Roman coin with an image of a ram looking back at a star, Dr Molnar began to investigate its possible meaning. In order to do so, he had to start thinking like a first-century astrologer rather than a twentieth-century astronomer. Likewise, as we think and pray about the possible meanings of the Star of Bethlehem, we too must walk in the sandals of the first-century people Matthew was writing to when he first told the star's story in his Gospel.

We will first go through a list of possible explanations for the star, asking whether it was supernatural or mythical, or a natural celestial event, such as a comet, a supernova or maybe a conjunction of several celestial bodies? I believe that in each case we can show how these twenty-first-century explanations do not fit what we know about the star and that we need to think more like someone from the last century BC or the first century AD.

Exploring the origins and development of astronomy will be helpful in understanding how people at the turn of the first millennium perceived the night sky. The night sky has not changed all that much over the last two thousand years, but how we interpret what we see is very different from how our ancestors perceived what they saw.

We will try to get ourselves into a first-century frame of mind by looking at the history that preceded the meeting of King Herod and the Magi. The so-called deuterocanonical or intertestamental period, the time between the return of the Israelites from Babylonian exile to the appearance of John the Baptist, spans about four hundred years. During this time, the scriptures are practically silent.

The prophetic voice may have been quiet, but even so, a lot was happening in the world, with perhaps the most consequential occurrence being Judaism coming into contact with Hellenism.

Hellenism – Greek culture – is a major factor we need to consider. Learning about the philosophies the Gentile peoples used to help organise their understanding of the world around them will help us to appreciate why belief in astrology appears to have been so widespread.

We will dabble in Hellenistic astrology. We will only be touching on this difficult and arcane topic so that we can identify what the Magi would have considered an iron-clad portent pointing to a regal birth in Judaea.

The story that we are entering into is convoluted, nuanced and confusing. I am sure that some of the confusion will be a result of my own shortcomings; for that, I ask up front for your forbearance, and I encourage the reader to persevere.

Matthew includes the star in his nativity story for a reason. The Star of Bethlehem is not an *ignis fatuus*, a will-o'-the-wisp. No less a scholar than Pope Benedict XVI believed in the historicity of Matthew's Gospel. He firmly holds that the first two chapters of the Gospel dedicated to the infancy narrative are not merely "a meditation presented under the guise of history". Pope Benedict insists: "Matthew is recounting real history, theologically thought through and interpreted, and thus he helps us to understand the mystery of Jesus more deeply."[4]

---

[4] Pope Benedict XVI, *Jesus of Nazareth: The Infancy Narratives* (New York: Image, 2012) p. 119.

Possible Explanations for the Star

D r Molnar begins his study by noting that there are four types of explanations for the Star of Bethlehem. These categories are supernatural, natural, mythical or astrological.

Messenger from God

Some have wondered if the star was a supernatural or miraculous event. In fact, could the star have been something like an angel?

Angels are often depicted in Scripture as messengers of God, intermediaries between heaven and earth. In the first chapter of Matthew's Gospel, when Joseph learns that his betrothed, Mary, is with child, he considers sending her away quietly. "But as he considered this, behold, an angel of the Lord appeared to him in a dream, saying, 'Joseph, son of David, do not fear to take Mary your wife, for that which is conceived in her is of the Holy Spirit'" (*Mt* 1:20). Later, after the birth of Jesus, Matthew tells us how, after the Magi departed, "an angel of the Lord appeared to Joseph in a dream and said, 'Rise, take the child and his mother, and flee to Egypt, and remain there till I tell you; for Herod is about to search for the child, to destroy him'" (*Mt* 2:13).

So, it is clear that Matthew is familiar with the role of angels as messengers and is apparently quite willing to use them to help move the plot line along. Even so, the Magi insist that it is a star, not an angel, that brought them to Jerusalem in search of the newborn king of the Jews: "For we have seen his star in the East, and have come to worship him" (*Mt* 2:2). Matthew does not say that the star is miraculous, let alone an angel; he just calls it a star and leaves it at that.

Comets, novae and conjunctions

Perhaps ever since astrology and astronomy parted ways, in the late 1500s to early 1600s, people have looked to visible natural celestial phenomena to explain the Star of Bethlehem. As moderns, we tend to think that the Star of Bethlehem must have been visually significant and so would demand people's attention. Though conjunctions can be beautiful, they are subtle and a bit of an acquired taste. It is not surprising, then, that comets and supernovae are often what head the list of possibilities.

Comets

Comets can be spectacular. I recall going out night after night to marvel at Comet Hyakutake during its apparition in the spring of 1996: truly a remarkable spectacle. The comet was easily seen by the naked eye as it stretched over a large portion of the night sky. This experience of a comet like Hyakutake was in sharp contrast to the let-down I felt at the 1986 apparition of Halley's Comet, which was nowhere near as impressive.

Comets, like the Star of Bethlehem, appear to drift about against the backdrop of the stars in the night sky. So, they check the box for a star that moves around in the sky. However, comets are often seen as bad omens (just ask King Harold II of England), which does not fit with the Magi's reaction: "When they saw the star, they rejoiced exceedingly with great joy" (*Mt* 2:10).

It would seem that a comet is an unlikely candidate for the Star of Bethlehem.

Halley's Comet as depicted on the Bayeux Tapestry. In 1066, the comet was seen in England as an omen. Later that year, England's King Harold II died at the Battle of Hastings when the Norman king, William the Conqueror, claimed the throne.

Supernovae: stars that go bump in the night

A star's life is a constant balance between the outward pressure of energy trying to escape the stellar core and the force of gravity trying to crush it. Stars form in large interstellar clouds composed of gas and dust. It is thought that a local gravitation instability, perhaps a little too much mass in one place, results in even more material being drawn in. As the clump's mass increases, so does its gravitational reach, which results in even more mass being drawn in. Skipping over a lot of important and interesting details such as conservation of angular momentum and accretion disks, the protostellar core heats up as the material falling in converts gravitational potential energy into heat. The density and pressure of the core increase with the protostar's ever-growing mass. If there is enough material within the protostar's gravitational reach, the temperature will become high enough to begin nuclear fusion. Nuclear fusion is the process by which lighter elements fuse together or are "synthesised" into heavier elements. Astronomers often refer to nuclear fusing of atoms as "burning", even though it is no way similar to the burning of a candle or fuel oil.

Initially, hydrogen atoms fuse to form helium, and a lot of excess energy is released in the process. Eventually, the gravitational pull inward is balanced by the outward pressure due to the energy being produced in the stellar core. This balancing act continues for as long as the star has "fuel" to "burn".

For stars that start life with five or more times the mass of our Sun, things start to get very interesting when the fuel begins running out. First, not only can these massive stars fuse hydrogen into helium, but they can also switch to fusing helium into carbon and switch again to fusing carbon into oxygen. At each transition, as the former fuel runs low and energy production in the core slackens, gravity once again tries to collapse the star. As gravity pulls the star in on itself, the density and pressure of the core increase, driving up the temperature. Eventually, the core temperature rises high enough to enable the next fuel source to begin the fusion process and a new equilibrium to be established.

However, one day, the star runs into a problem when the silicon atoms it has been synthesising into iron begin to run out. The problem is that iron nuclei are so tightly bound that trying to add more neutrons or protons causes the nuclei to break apart. Nuclear fusion ends with iron, and with it the outward flow of energy that has been offsetting gravity's inward pull. The star is out of fuel. In massive stars, the force of gravity can become so great that a new reaction begins: the protons and electrons that make up the iron in the core merge into neutrons. Within a second, the roughly Earth-sized iron core collapses into a sphere of neutrons, only about 10 km in diameter.

Suddenly, the outer layers of the star have no support. Like a building whose lower floors have been blown out, the star's outer layers come crashing down onto the neutron core. The resulting cataclysmic explosion is called a supernova. In the minutes following core collapse, the star can radiate more energy than it did during its lifetime before that moment. The luminosity of a star going supernova can momentarily be greater than that of all the other stars in the galaxy combined. If a supernova happens to occur close enough to Earth, it can be bright enough to be seen by the naked eye. These "guest stars" appear in the night without warning, shine for a while, and then fade away. Depending on the initial mass of the star, its corpse will either be a neutron star or, if it was hefty enough, not even the nuclear forces between neutrons can stop the star from collapsing into a black hole.

The remnants of supernova explosions litter our galaxy. Astronomers can analyse them and have a fair idea of when the explosion occurred. According to Eric Betz in the November 2020 issue of the magazine *Astronomy*,[5] there are seven naked-eye supernovae in the historical record for which astronomers can identify a likely stellar corpse. The earliest recorded guest star, Supernova RCW 86, was first observed in the constellation Circinus by Chinese astronomers in AD 185.[6] However, since we believe that

---

[5] https://www.astronomy.com/science/7-naked-eye-supernovae-throughout-human-history/

[6] What we have just described is a "core collapse" or Type II supernova. As you might infer, there is also a Type 1. A Type I supernova occurs when a white dwarf star (the stellar corpse from a star like our Sun) accumulates sufficient material (perhaps from a binary companion orbiting a bit too close), igniting runaway carbon fusion, which disrupts (destroys) the dwarf star. The AD 185 supernova appears to have been a Type I.

Jesus was born 185 years earlier, give or take a few years, it is unlikely that this guest star was the Star of Bethlehem.

There are hints in Chinese records of another guest star appearing in 134 BC, in the constellation Scorpius. This same supernova might also be referenced by Hipparchus. Again, for our purposes, this supernova has a timing problem: it occurred long before our earliest estimated dates for Jesus's birth.

Supernova RCW 86 was observed towards the southern constellation of Circinus in AD 185 by Chinese astronomers. This is currently the earliest known naked-eye supernova for which supernova remnants have been identified by modern astronomers.

Someday, astronomers studying supernova remnants may stumble across one that is close enough to Earth to have been visible to the naked eye and whose timing falls within our estimates for the birth of Jesus. Till then, the data does not seem to support the hypothesis that the Star of Bethlehem was a supernova.

Conjunctions

Conjunctions are the clustering of two or more celestial bodies in close proximity as they appear in the sky. If one object actually passes in front of another it is called an occultation. If the objects involved in the occultation are the Sun and the Moon, we call it an eclipse.

Some have suggested that a conjunction of planets, such as Jupiter and Saturn, was the basis for the star. For instance, the famous astronomer, Johannes Kepler, himself known to have cast a horoscope or two in his day, suggested that a conjunction of Jupiter and Saturn in 7 BC may have been the star. He also proposed, after he witnessed a new star, a supernova, appearing near a similar conjunction of Jupiter and Saturn in AD 1604, that perhaps a similar supernova was spawned by the 7 BC conjunction. This supernova, he opined, may have been the Star of Bethlehem. We know today that supernovae are not the result of conjunctions between planets. They are, in fact, explosions caused by certain types of stars at the end of their lives. However, we should congratulate Kepler for managing to combine two popular explanations for the Star of Bethlehem into one.

Though conjunctions and occultations can be quite beautiful, they are nowhere near as flashy as a good comet or a nearby supernova. They are also not particularly rare.

Mythical

If the star was neither natural nor supernatural, then maybe it never existed. Perhaps the star is unhistorical. It has even been suggested that Matthew could have added the star to make the birth of Jesus seem more important.

This last-mentioned conjecture is always a possibility, of course, but unless documentation should turn up in which Matthew admits to inventing the Star of Bethlehem, this conjecture cannot be proven. Furthermore, I am unaware of the evangelist being shown to have engaged in any kind of deception, and so we have no reason to doubt what he wrote. As theologian and New Testament scholar Klaus Berger says:

> Even when there is only a single attestation... one must suppose, until the contrary is proven, that the evangelists did not intend to deceive their readers, but rather to inform them concerning historical events... [T]o contest the historicity of the account on mere suspicion exceeds every imaginable competence of historians.[7]

---

[7] As quoted in Pope Benedict XVI, *Jesus of Nazareth: The Infancy Narratives* (New York: Image, 2012) p. 119.

Astrological

So, there is one other possibility left to be considered. We know that the ancients were interested in conjunctions and occultations, not as natural events but as astrological omens. We are left to consider that the Star of Bethlehem was an astrologically significant apparition – a portent of a regal birth.

First, though, let us find out more about the history that formed the people present at the meeting of the Magi with King Herod.

From Babylonian Exile to News of a Newborn King

It has been noted that when Malachi, the last "writing prophet", laid down his pen in the middle of the fifth century BC a hush fell on the prophetic voice. This hush had lasted nearly five hundred years by the time John the Baptist suddenly appeared in the desert, crying out "Prepare the way of the Lord." It would seem that from Malachi to John the Baptist not much happened.

Nothing could be further from the truth.

Many things happened, and not just in terms of empires conquering one another. The distances over which people could travel and communicate had expanded. For example, the travel distance today by car and ferry from Athens to Jerusalem is around 2,100 km, while from Rome to Jerusalem it is about 3,500 km (both routes include tolls). We might find it daunting to drive these distances, and yet we know that the people of the first millennium were travelling between these cities regularly. Not everyone was, of course, mostly the rich and the powerful. Yet, even so, the world was becoming smaller. A side benefit was that knowledge and ideas developed by one culture could now spread far and wide, influencing other cultures and making it less likely that the knowledge could be lost entirely in the case of a local calamity.

Almost a hundred years before Malachi picked up his pen, Israel's Babylonian exile had ended when the up-and-coming Achaemenid Persian Empire overthrew Babylon. The Persian leader, Cyrus the Great, decreed that the Jews could return to their homeland. Over the next hundred years, waves of exiles returned to Judaea. Zerubbabel led the first wave around 537 BC and set to the task of rebuilding the Temple. Circa 458 BC, Ezra the Scribe arrived to reinvigorate the spiritual life of the people. Soon after, Nehemiah came to rebuild the city walls to ensure the people's physical safety.

Around about this time, Malachi points out that while the return of the Jewish people to their homeland and their rebuilding of their Temple are all well and good, there are still some problems needing to be addressed. For instance, the sacrifices being offered are not of good quality (*Ml* 1:6-9), people are failing to tithe (*Ml* 3:8-10) and divorce is being tolerated (*Ml* 2:13-16).

Malachi also anticipates the coming of a Messiah. He tells of a herald who will prepare the Messiah's way: "See, I am sending my messenger to prepare the way before me, and the Lord whom you seek will suddenly come to his temple. The messenger of the covenant in whom you delight – indeed, he is coming, says the Lord of hosts" (*Ml* 3:1).

Then, all seems to suddenly go quiet. The written record ends. The prophetic voice goes mute.

When we finish reading the Hebrew Scriptures and then pick up the New Testament, it is as if the Jewish people return from the Babylonian exile and, suddenly, it is four hundred years later. The Magi from the east are arriving in Jerusalem asking King Herod to kindly direct them to the newborn king of the Jews. However, a whole lot of interesting things have happened in those intervening years that have a direct impact on that meeting between the Wise Men and Herod.

For starters, just as the Jewish people were picking up the pieces of their shattered Temple and society, an "earthquake" had occurred that had altered the political, cultural and intellectual landscape of the Near East. At the epicentre of this earthquake was a man: Alexander the Great.

Greece

For a large part of Israel's history, Egypt was the dominant power in the region. Then the Assyrians became the major player. When the Assyrian Empire fell, it was Babylon who took its place. The Persians overthrew Babylon. Next, Greeks appeared on the scene.

The Greeks had a long history of conflict with the Persians. As the Achaemenid Persian Empire expanded westward they first bumped into regions along the eastern Aegean shores inhabited by Greeks and then the Greek city states themselves. In 547 BC, Persian leader Cyrus the Great (who at this very same time was allowing the Jews to return to their homeland) conquered one of these regions, Ionia. Ionia was located on the west coast of Anatolia (Asia Minor) in the area of the present-day Turkish city of İzmir. As new subjects of the Persian Empire, the inhabitants of Ionia were allowed "self-rule" under Greek tyrants appointed by the Empire. So long as the Ionians were docile and paid their taxes, the Persians were content to let them be. However, the independent-minded Greeks were anything but docile.

One of these tyrants, Aristagoras, leader of the Ionian city of Miletus, incited a revolt against the Persians in 499 BC. Much of Hellenic (Greek) Asia Minor followed Ionia into rebellion. For the next fifty years (c.499-449 BC, the same time that the Jews were rebuilding their society) there were a series of conflicts referred to as the Persian Wars.

Eventually, the Persians decided that the Greeks were a serious distraction if not an outright threat to the Empire and needed to be dealt with decisively. Darius the Great began planning an invasion to conquer Greece, but died in 486 BC before he could mount the invasion.

He was followed on the throne by his son Xerxes. In 480 BC, Xerxes personally led the planned invasion, crossing into Greece with what is thought to be one of the largest armies assembled up to that time. The Persians were victorious at the Battle of Thermopylae. They overran much of Greece and burnt Athens. However, the Fates, who had been favouring the Persians up to this point, started favouring the Greeks, who managed to defeat the Persian fleet at the Battle of Salamis. A year later, the combined Greek forces defeated the Persians at the Battle of Plataea, ending the second Persian invasion.

Further battles between the Greek city states and the Achaemenid Empire occurred, but by the middle of the fifth century BC things had quietened down. The Persians appeared to change tactics. Instead of direct warfare, the Persians now maintained a strategy of keeping the fractious Greek city states at one another's throats. The plan was this: keep the Greeks distracted with their own internal fighting, which would prevent them from turning their attention to Persians and, thus, keep the western borders of the Persian Empire calm. The strategy worked. There was relative quiet. But nothing lasts forever. The Greeks now saw the Persians as an existential threat to their culture and decided that they needed to go on the offensive.

Alexander the Great

About a hundred years after things quietened down, King Philip II of Macedon (359-336 BC) managed to consolidate much of classical Greece into a federation known as the League of Corinth (338/7 BC). He also made changes to the military that greatly increased its effectiveness on the battlefield. With political consolidation and an improved military, Philip II was ready to invade the Achaemenid Persian Empire.

History does not repeat itself but often it rhymes. Recall how it was the Persian king, Darius the Great, who constructed a great army to invade and crush the Greeks, but how, upon his death, it instead fell to his son Xerxes to execute the plan.

In 336 BC, it was Philip II who died, assassinated by a member of the royal bodyguard before he could launch an invasion of Persia. Not surprisingly, Persian involvement was suspected. If it were the case that the Persians engineered Philip's assassination, they were doubtless convinced that whoever replaced him would be inferior. They were to be gravely disappointed.

The planned invasion of Persia fell to Philip's son, Alexander. After consolidating his position as King by calming revolts in places such as Thebes and Athens, Alexander was given the title, "Hegemon of the Hellenic League", the leader of the Greek forces against the Persians.

In 334 BC, at the age of twenty, Alexander crossed the Hellespont, passing from Europe into Asia. In 331 BC, he defeated Darius III at the Battle of Gaugamela. His army marched to the Persian ceremonial capital, Persepolis. For reasons that are now lost to us, the Persian king's palaces burned. Some suggest that it was retaliation for the burning of Athens during the Persian wars. Others believe too much wine may have played a part in the decision to destroy the city. In any event, it is said that Alexander eventually regretted the destruction of Persepolis.

Alexander died only seven years later (323 BC) in Babylon at the age of thirty-two, but only after having been named Pharoah of Egypt and King of Persia.

Alexander was great at many things, but apparently succession planning was not one of them. Now, to be fair, his death came suddenly and, at the age of thirty-two, unexpectedly. Ultimately, his empire, stretching from Macedon to the Indian subcontinent, was divided amongst his generals. Two of these generals are of particular interest to us here: Ptolemy and Seleucus. Ptolemy's dynasty would rule Egypt, with Alexandria their capital city. Seleucus and his successors would rule over Mesopotamia and Syria, with Antioch as their capital.[8] Unfortunately for the Jews, they once again found themselves sandwiched between two competing empires. Through much of their history, the Jews have found themselves on the frontiers of an empire of the Nile and an empire of the Euphrates. As with many border regions, the exact location of the borders was often in dispute.

Hellenic hegemony

Up till now, whenever the Jews were under the influence or control of this or that empire, we could point to a specific ruler of that empire and to the empire's capital city. In addition, moving from the control of one empire to another would also mean a significant change in culture. Under

---

8 Anderson, Bernhard W., *Understanding the Old Testament,* 4th ed. (Englewood Cliffs, NJ: Prentice Hall, 1986) p. 612.

the Greeks, though, things were different. Mind you, it is still important to know *which* Hellenistic kingdom was dominating Jerusalem: the question became whether the Jews were dealing with a *laissez-faire* or a muscular form of Hellenism.

Hellenism is a term used to describe the attributes of Greek culture that followed in the wake of Alexander's conquests and were cultivated by his successors. Alexander was not motivated merely by the thrill of military conquest but also by a desire to spread the best of Greek culture. Remember, Alexander had none other than Aristotle as one of his tutors. His education included logic, rhetoric, music and astronomy. Alexander and his generals felt it a duty to "evangelise" the glories and the benefits that flowed from Greek culture. Even while the Hellenistic kingdoms jockeyed amongst themselves for political, economic and military dominance, they all continued to work towards Alexander's goal of a world brought together by Greek culture.[9]

After Alexander

Initially, it was the Ptolemies in Egypt who ruled over the Jews. It is said that their rule out of their capital in Alexandria was gentler than the more aggressive stance taken by the Seleucids in Antioch.[10] Both kingdoms competed to be cultural centres that might rival even Athens. In this arena,

9 *Ibid.*, p. 612.

10 *Ibid.*, p. 613.

Alexandria seems to have outpaced Antioch. The loss of the Museum of Alexandria with its library, that famed centre of arts and sciences, can still bring antiquarians to tears when they think of all that was lost to fire and time.

Jerusalem was under the control of the Ptolemies during the third century BC. It was around this time that work began on the Septuagint, the translation of the Hebrew Scriptures into Greek. The mere fact that such an effort was undertaken suggests that there was a significant population of Greek-speaking Jews and Gentiles who were interested in the Hebrew Scriptures. Since Greek was a *lingua franca* of the Roman Empire, the Septuagint would also be used by early Christians.

Meanwhile, the Seleucids were trying to wrest control of the Mediterranean coastlands from the Ptolemies. The Seleucids argued that this region was rightfully theirs, which would put Jerusalem and Judaea under Syrian rule. In 223 BC the political geography started to change when Antiochus III (the Great) came to power in Syria. He began a war with the Ptolemies in Egypt that would go on for over twenty years.

In 198 BC, Antiochus the Great defeated Ptolemy V at Paneas, (later Caesarea Philippi, located close to the source of the Jordan river). This finally brought Judaea under Syrian control.[11]

---

[11] *Ibid.*, p. 615.

As mentioned above, the Seleucids tended towards a more muscular brand of Hellenism. If Antiochus the Great was a champion of all things Greek, then his successor, Antiochus IV, was fanatical. Pressure mounted on the Jewish people to let go of those things that separated them from the Gentiles and embrace Greek culture. Antiochus IV's zeal for all things Greek would push his new subjects to breaking point.

Maccabean revolt

Wars of conquest do not come cheap. Antiochus the Great finally achieved the goal of wrestling the Mediterranean coastlands from the Ptolemies, but he did so by draining his treasury. His successor, Antiochus IV, needed to replenish the coffers, and so he increased taxes, which did not endear him to his new Jewish subjects. When Antiochus began selling the office of High Priest to raise more funds there was rioting. Things went from bad to worse when he used his army to try and subdue the restive Jews while also taking the opportunity to plunder the Temple of its treasures.

The Jews stubbornly continued their defiance. Ultimately, Antiochus outlawed the Jewish religion altogether and demanded the complete Hellenisation of Jewish life. It was at this point that acts of resistance by members of a Jewish family sparked a full-scale insurrection. This family and the people who flocked to their cause would become known as the Maccabees.

The Maccabean Revolt, 167-160 BC, achieved the unimaginable: almost a century of self-determination for the Jewish people. No doubt, geopolitical changes favouring the Jews in their struggle for freedom were already occurring. For one thing, there was the rise of a new world power – Rome. Nonetheless, for the first time since the Kingdom of Judah had been defeated and led into exile by the Babylonians, the Jewish people had won for themselves almost a century of self-determination.

After they deposed their Syrian rulers, the Maccabees established the Hasmonean dynasty. This would govern the Jewish people during their period of independence. But, alas, following an all-too-familiar pattern, what began so brightly soon began to fade. For one thing, the Hasmonean leaders always had about them a whiff of illegitimacy since they were not of the Davidic line. They then appropriated the office of High Priest and begin to sell the office to the highest bidder. Finally, over time, the Hasmonean kings and queen became more and more influenced by Greek culture. Again, history does not repeat itself, but often it rhymes.

As noted earlier, Rome's increasing interventions in the region may have aided the Jews in achieving independence by distracting Alexandria and Antioch. If this was indeed the case, then it is certainly ironic that it would be a Roman general, Pompey, who would bring Jewish self-determination to an end when he entered Jerusalem in 63 BC.

Pharisees, Sadducees and the Hasmonean Civil War

While I was a seminarian, one of my professors, Fr James Massa, explained the difference between the Pharisees and Sadducees. To be honest, I never considered there to be much in it. They are often paired together in the Gospels, and both gave Jesus a hard time. As far as I was concerned, one was pretty much the same as the other.

Well, Fr Massa explained that one of the big differences had to do with their beliefs about the afterlife. Christians believe that, at the end of time, when Jesus comes again in glory, our bodies will be raised from the dust of the earth and reunited with our souls. We testify to this belief when we say at the end of the Creed "I look forward to the resurrection of the dead, and the life of the world to come." Fr Massa explained that, like Christians, the Pharisees believed in the resurrection of the body. The Sadducees, on the other hand, did not. As far as the Sadducees were concerned, "When you're dead, you're dead, and that's it." As Fr Massa humorously put it, "that is why they were *sad, you see*".

Contrasting beliefs about the afterlife was a major difference between Pharisees and Sadducees, but there were others, including their differing ideas about Hellenism and its role in Jewish life.

The Sadducees came to be during the time of Hasmonean rule. They were the group who favoured Hellenisation[12]

---

[12] Cavins, Christmyer and Gray, *The Bible Timeline: The Story of Salvation* (West Chester PA: Ascension, 2019) p. 129.

and advocated for a policy of tolerance and compromise.[13] They tended to come from the upper levels of society, the priestly and political classes – just the sort of people who, through their educational and commercial activities, would be expected to come in contact with Hellenism.[14] However, though they were open to Hellenism, they were strict followers of the written word in the Torah (the first five books of the Hebrew Scripture), and they rejected any oral tradition. It was because of their strict adherence to the written word that they rejected the doctrine of resurrection; they could not find any evidence for it in the Torah. Similarly, they rejected belief in angels and demons as well as prediction about the end time.[15]

The other major group were the Pharisees. They, too, revered the Torah; however, they were more liberal than the Sadducees because they also accepted other books such as the Prophets, as well as oral tradition. Unlike the Sadducees, the Pharisees were extremely wary of external influences such as Hellenism. They placed emphasis on those customs that maintained a separation of Jews from Gentiles, such as circumcision and dietary rules.[16] Compromise with Hellenism was a non-starter for the Pharisees.

---

[13] Anderson, Bernhard W., *Understanding the Old Testament* 4th ed. (Englewood Cliffs, NJ: Prentice Hall, 1986) p. 634.

[14] *Ibid.*, p. 613.

[15] *Ibid.*, p. 634.

[16] *Ibid.*, p. 634.

In time, the tensions between these two groups worsened, and when this was coupled with a succession struggle within the Hasmonean dynasty, the result was open conflict – a civil war.

What happened was this. In 67 BC, Queen Salome Alexandra (it is, perhaps, ironic that the Hasmonaean ruler should be named Alexandra, given that it is the feminine form of the given name Alexander), the one and only queen of the Jews, died. She left two sons, Hyrcanus and Aristobulus. Hyrcanus, the older of the two, appears to have favoured the Pharisees, while Aristobulus was more in line with the Sadducees.

Two years after their mother's death, the forces of the two brothers battled at Jericho. Unfortunately for Hyrcanus II, many of his soldiers defected over to his brother's side. Negotiations ensued where it was agreed that Aristobulus should succeed his mother and become king and Hyrcanus quietly become a private man.[17]

The agreement did not last.

A very rich adviser to Hyrcanus II, a Jewish Idumean by the name of Antipater, encouraged Hyrcanus to take back the throne.[18] Hyrcanus hesitated but then agreed. Antipater reconstituted Hyrcanus's army with aid from the Nabataeans,

---

[17] All translations of Josephus's *Antiquitates judaicae* follow the translation in The Works of Josephus; The Antiquities of the Jews, trans. William Whiston, (Peabody MA: Hendrickson, 1987) *A.J.* 14.4-7.

[18] *Ibid. A.J.* 14.8.

the people of an Arab kingdom whose capital, Petra, was south of Jerusalem. With combined Arabian and Jewish forces, Aristobulus's army was defeated, and Aristobulus II fled to Jerusalem and took refuge in the Temple area. Hyrcanus's army laid siege to the Temple at the Passover of that year.[19] At that point, each of the brothers reached out to the Romans for support.

Rome

As noted earlier, since the beginnings of the Hasmonean dynasty, Rome had been in the process of becoming more active in the region. The rise of Roman influence in the region continued through the decades following Judaea's independence. Rome was also in the process of transforming from a republic to an empire. By 64 BC, the Roman general Pompey was consolidating a new Roman province out of what had been Syria. In their fight between each other, both Hyrcanus II and Aristobulus II sent messengers to Pompey asking for his backing. Pompey intervened on the side of Hyrcanus II, and Aristobulus II was eventually convinced to step aside, though many of his followers did not surrender, but instead took refuge in the Temple.

After several months of preparation, and using siege engines, the Romans broke through the Temple defences and killed many of the defenders and priests. Pompey established Hyrcanus II as high priest but did not make him

---

[19] *Ibid. A.J.* 14.19-21.

King. Eventually, Hyrcanus II is named an ethnarch: a title used to describe leaders of a common ethnic group. Judaea remained autonomous, but was now dependent on Roman administration based in, of all places, Syria. Jewish self-determination had effectively come to an end.

Reflecting on how the Jews once again found themselves ruled from Syria, Flavius Josephus, the first-century Jewish-Roman historian laments:

> Now the occasions of this misery which came upon Jerusalem where Hyrcanus and Aristobulus, by raising a sedition one against the other; for now, we lost our liberty, and became subject to the Romans, and were deprived of that country which we had gained by our arms from the Syrians, and were compelled to restore it to the Syrians.[20]

King Herod the Great

The reader will recall the Jewish Idumean, Antipater, the rich friend who convinced Hyrcanus II to reconsider abdicating. One of Antipater's sons, named Herod, became a close friend of Mark Antony. Yes, that is right: Mark Antony of Cleopatra fame. Herod also became the king of Judaea.

Antipater's support of the Roman general Pompey began a long association between his family and Rome. In 47 BC, Julius Caesar named Antipater procurator of Judaea. Antipater, in turn, named his son Herod governor

---

[20] *Ibid. A.J.* 14.77.

of Galilee. Six or so years later, his above-mentioned friend, Mark Antony, made him tetrarch of Galilee.

In 40 BC, the Parthians invaded Syria and Palestine. The Parthian Empire was located east of the Euphrates and, at its greatest extent, stretched from today's central Turkey to Afghanistan to western Pakistan. The Parthians and the Roman Republic (and later the Roman Empire) engaged in a series of conflicts along their common frontiers referred to as the Roman-Parthian Wars (54 BC-AD 217). In a nutshell, during the 40 BC invasion, the Parthians took Jerusalem; Herod's brother was taken captive and committed suicide; Antigonus, from the old Hasmonean dynasty, was installed as ruler in Jerusalem; and Herod fled to Rome.

The Roman Senate, who were not amused by Parthenian incursion into Roman territory, declared Herod King of Judaea and sent him back home with an army. By 37 BC, Herod had established practical control in Jerusalem. Once he established control, the reign of Herod I was relatively peaceful. He managed, by being an able administrator who was murderously brutal, to remain in power for some thirty-two years. Administratively, he oversaw the construction of the palace fortress Herodium and the city of Caesarea Maritima. With an audacity that might even make a Caesar blush, Herod doubled the area of the Temple Mount to make it big enough to accommodate his vision for remodelling the Temple. Herod's Second Jewish Temple replaced the one built by Zerubbabel and the Jews upon their return from

Babylon. Josephus saw Herod's Temple before its destruction in AD 70 and describes its magnificence in this way:

> Now the outward face of the temple in its front wanted nothing that was likely to surprise either men's minds or their eyes, for it was covered all over with plates of gold of great weight, and, at the first rising of the Sun, reflected back a very fiery splendour, and made those who forced themselves to look upon it to turn their eyes away, just as they would have done at the Sun's own rays. But this temple appeared to strangers, when they were at a distance, like a mountain covered with snow; for, as to those parts of it that were not gilt, they were exceeding white.[21]

Herod also maintained power by being brutal. He did not hesitate to kill anyone – family or friend – whom he perceived as a danger to his rule. He killed his wife, Mariamne, on suspicion of being unfaithful. Several of his sons were executed for disloyalty, real or perhaps imagined. His firstborn son, Antipater II, was executed only five days prior to Herod's own death in 4 BC. Famously, Augustus Caesar is said to have quipped, "It is better to be Herod's pig than son" (Macrobius, *Saturnalia*, 2:4:11).[22] It should be pointed out that it is said that Emperor Augustus was a friend

---

[21] The Works of Josephus; The Wars of the Jews, trans. William Whiston, (Peabody MA: Hendrickson, 1987) *B.J.* 14.4-7.

[22] https://www.catholic.com/magazine/online-edition/it-is-better-to-be-herods-pig-than-son

of Herod, which suggests that even his friends understood how ruthless a man he was.

Alongside his major building projects, Herod tried to raise Judaea's standing in the Hellenistic world of the Mediterranean by making extravagant gifts to Athens and being a patron to the Olympic Games. He also built a theatre and an amphitheatre in Jerusalem, both trademark Hellenist cultural institutions. Taken all together, although Herod always identified as a Jew, he also appreciated Hellenism.

And so, some three decades after first taking power, and only a few years before his death, King Herod the Great met with the Magi from the east. At this meeting, Herod would become frightened, and all Jerusalem with him, when the Hellenist astrologers asked of him, "Where is he who has been born king of the Jews? For we have seen his star in the East and have come to worship him" (*Mt* 2:2).

Astronomy, Astrometry and Astrology

The huge dome of the sky is of all things sensuously perceived the most like infinity. And when God made space and worlds that move in space, and clothed our world with air, and gave us such eyes and such imaginations as those we have, He knew what the sky would mean to us. And since nothing in His work is accidental, if He knew, He intended.

C.S. Lewis, *Miracles*

All my life, I have enjoyed spending time under a clear night sky watching the dance of the planets, trying to spy a naked-eye comet, or hunting some elusive deep-sky object with a telescope. I have spent a lot of time outside at night looking up at the sky; indeed, it has become a habit of mine to look up at the sky anytime I step outside at night.

One warm and clear autumn night, my astronomy-graduate-student friends and I went to an observatory located at nearby Amherst College. For those who know their telescopes, the Wilder Observatory telescope is an 18-inch (46 cm) Alvan Clark refractor. Around 24 feet long

(7.3 m), this telescope is a splendid example of President Teddy Roosevelt-era high-tech, and we had the keys to the observatory. So, when we probably should have been studying for exams, we were instead "joy riding" this venerable telescope all around the night sky.

It was very late, probably two or three in the morning, when we finally put the old telescope back into its stow position, closed the observatory dome's shutters and exited the building by a door that happens to face towards the east. Automatically, I looked up as I stepped outside, and there, rising over the trees in the southeast, was my old friend, Orion the Hunter. I say "my old friend" since a good portion of my PhD research targeted the large molecular cloud (M42) located in the Orion constellation. For the previous few winters, when Orion was best positioned to be observed from New England, I had spent many hours at the Five College Radio Observatory making observations for my doctoral thesis.

As I mentioned, it was warm for an autumn night in New England, but when I glanced up and saw Orion, I do not exaggerate when I say that the air temperature instantly dropped twenty degrees Fahrenheit (eleven degrees Celsius). Well, it seemed to anyway. I remember thinking, "Now I understand how strong the associations must have been for our ancestors." When Orion first appeared in the eastern sky, they knew that the storms and cold of winter were soon to follow. As a matter of course, our ancestors were more attuned to the night sky than we are today. If they were to be asked what the phase of the Moon was, they could probably give the correct answer. If I were to be asked, I would probably go to the internet to find out.

I relate this story to try and drive home just how easy it is to make associations between the patterns we see in the heavens and the patterns of our lives on Earth. We humans seem hardwired to *see* connections between events, even if we often fail to discern whether these connections are causal or merely correlations. Be that as it may, our ancestors viewed celestial phenomena as possible omens and portents of what the future may hold. What motivated the scribes in ancient Mesopotamia to begin the development of mathematical astronomy was not just a desire to know the present time of day or season of the year but also a desire to divine knowledge of future events, both auspicious and inauspicious. Not surprisingly, political and military leaders, who in every age are looking for an advantage over their competitors, were keenly interested in what the scribes could divine from their astronomical observations. Kings, in particular, were interested in what the stars might have to say about human affairs and terrestrial events, especially when the portent might relate to their successor.

An astronomy primer

Before we start delving into astrology and just what the peoples of the first century BC would have considered an iron-clad portent of a regal birth, let us take some time to learn or brush up on some basic astronomy. Please keep in mind that many of the concepts presented here are not newly discovered but are actually quite old. These concepts are how our ancestors first started making sense of the sky. The journey they began, thousands of years ago, has brought us to our present-day understanding that the apparent motions of celestial objects in the sky are primarily due to Earth's own rotational motion about its axis, coupled with its orbital motion around the Sun. But we are jumping ahead of the story. For astronomers of the first century, their working model of the sky was that of nested spheres to which the various classes of celestial objects were attached.

One of the challenges we face in understanding the Star of Bethlehem's meaning is to first understand how the celestial portent was perceived by ancient peoples. We need to set aside our space-age understanding of the universe and put ourselves in the sandals of people of the Near East at the start of the first millennium. We need to see the sky as they viewed celestial events. This brief study of how early astronomy developed will give us a "feel" for the mindset of those early stargazers.

Even so, I will be using our modern concepts of Earth revolving on its axis while orbiting the Sun in order to explain what we see in the sky. Having these astronomical concepts and related terminology at our command will also aid in understanding the nature of the Star. Just keep in mind, though, that the ancient astronomers were still trying to figure all this out.

One final note: one thing the ancients did know is that Earth is a sphere. Their estimated value of Earth's diameter is very close to the value we know today. Chalk one up for our ancestors.

Where are we?

We live on a planet called Earth. The period of time it takes Earth to rotate once about its axis we call a day. Earth is the third of eight planets (nine if you count Pluto) orbiting a star called Sol. Earth completes one orbit around Sol every 365 ¼ days, or one year. Sol itself, also called the Sun, is an average-sized star, one of several hundred billion stars that comprise a large spiral system called the Milky Way galaxy. The Sun, along with all the other objects in our solar system, orbits the centre of the Milky Way every 230 million years, give or take.

The celestial sphere

Standing outside on a clear night looking up at the sky, we see a portion of the universe spanning over us like a dome. If our horizon is unobstructed, we can essentially gaze out at half the universe. Some of the celestial objects we see, such as the Sun, the Moon, the planets and the occasional comet, belong to our solar system. Objects belong to or are part of our solar system if they are "gravitationally bound" to the Sun – that is, wherever the Sun goes, these objects get dragged along with it by the force of the Sun's gravity.

The rest of the objects we see in the night sky are stars. A star is a large, massive, self-luminous spheroidal body of gas held together by its own gravity while being supported by the outward flow of energy being produced in its core through nuclear fusion. The stars we see in the night sky are not part of our solar system but are part of our Milky Way galaxy.

There is one more object that we might be able to see – if we have good vision and a dark sky – an object that is neither part of our solar system nor part of the Milky Way galaxy. This smudge of light in the constellation Andromeda is another galaxy. The Andromeda galaxy is a large spiral galaxy similar to our own Milky Way and is one of the most distant objects visible to the unaided eye.[23]

The distance to any of these objects is so large that we are unable to judge, using just our eyes, their true positions in space. In fact, they all appear to be at the same distance from us, as if they are attached to a very large dome spreading over our heads. We can imagine this dome being extended into what is called the "celestial sphere" with Earth at its centre. The whole visible universe, Sun, Moon, planets, stars and galaxies, can be imagined as being affixed to this imaginary celestial sphere. The celestial sphere does not have any physical reality, but it is what we call a "model" that helps us

---

[23] Unaided or so-called 'naked-eye' observations are astronomical observations made without the use of optical systems such as binoculars or telescopes; however, corrective optics such as glasses are allowed.

organise our observations. The Celestial Sphere model will aid us in thinking about the arrangement of the objects we see in the sky and their apparent motions.

Constellations

Now that we have imagined the whole universe as being stuck to the inside of an imaginary sphere of immense size, with Earth at its centre, how can we start to make sense of all these points of light that seem to be randomly spread over its surface? Estimates of the number of stars that are visible on any given night to the unaided human eye are in the range of anywhere from 2,500 to 5,000. Over the entire Celestial Sphere, anywhere from 5,000 to 10,000 stars are thought to be visible. So where do we begin to try and impose some order on this apparent chaos? One way to start is by noting the brighter stars in a patch of the sky and then "connecting the dots" to form a pattern. We then name the pattern after an object or creature, real or mythical. Sometimes, a pattern may resemble a figure, such as a charging bull, a dog or a man. Often, though, a lot of imagination may be needed to make the pattern match its name. The point is that grouping stars together and giving that grouping a name allows us to break the celestial sphere into more manageable "patches". It is an aid to memory, a mnemonic.

We call these patterns "constellations". Some of them actually do look a bit like their namesake. The constellation Taurus the Bull, for example, with some imagination (well, maybe a lot of imagination), can look like a charging bull. Orion does look like a man wearing a belt with one arm extended holding a bow and the other arm raised above his head. Canis Major has a vague resemblance to a stick-figure dog if you look closely.

Other constellations are, shall we say, more abstract. Draco the Dragon is a meandering string of stars, which might outline the snake-like body of a dragon, but your imagination will be needed to supply the wings. Another abstract constellation is Sagittarius, the Archer. I can only find this constellation if I imagine I am looking for a teapot. Perhaps the ancients called it an archer because teapots had not yet been invented. And then there is the constellation Lepus, which does not look much like anything, never mind a rabbit.

The usefulness of these memory aids in helping us find our way around the celestial sphere is improved if we can create stories that tie a group of constellations together into a narrative or tableau. For instance, when I look up at my old friend Orion the Hunter, I see him with his bow ready and club raised as he bravely faces the charging bull, Taurus. I cannot then help but note how his faithful hunting dog, Canis Major, is right behind him. That is, Canis is keeping Orion between himself and the ferocious charging Taurus; I guess this goes to show once again that canines are indeed intelligent creatures. As intelligent as he may be, Canis is nevertheless being distracted, by Lepus the rabbit, who is near Orion's feet. With this simple mnemonic, a story that binds together these four constellations, I now have command over a fair portion of the winter night sky.

Daily motion

Each day, the stars rise in the east and set in the west, just as the Sun does. The constellations, too, rise and set as a group across the night sky. This daily or diurnal motion is due to Earth's rotation. For the ancients, though, it seemed more plausible that it was the celestial sphere that was rotating rather than they who were moving. Let us not be too hard on our ancestors. Although the evidence favouring Earth's rotation would pile on more and more as time went on, the first direct observation of Earth's rotation did not occur until 1851 when Léon Foucault set his pendulum swinging beneath the dome of the Pantheon in Paris. As Foucault's pendulum swung, a stylus protruding from the bottom of the bob drew long ellipses in the sand that had been cast on the floor. These tracings in the sand seemed to shift clockwise over time as though the pendulum's swing was somehow rotating. In reality, it was Earth that was rotating the floor beneath a stationary pendulum.[24]

As the celestial sphere appears to rotate, carrying the stars on their courses around the sky, there are two points in the sky that do not move. In fact, it looks as if the whole

[24] Aczel, Amir D., *Pendulum: Léon Foucault and the Triumph of Science* (New York: Atria Books, 2003) p. 154.

celestial sphere pivots on these two points. These stationary points are the north and south celestial poles. The celestial poles are in line with Earth's rotational axis – that is, they are located directly above Earth's north and south poles. In the southern hemisphere, the south celestial pole has no visible marker; however, in the northern hemisphere, a star called Polaris, of the constellation Ursa Minor (the small bear), aligns almost perfectly with the north celestial pole. Commonly called the North Star or the Pole Star, Polaris is easily visible to the naked eye. Combining its easy visibility with a nearly stationary position over the North Pole makes Polaris a useful navigational aid for finding true north.

Along with the celestial poles, there is another useful marker that helps us navigate the celestial sphere, called the celestial equator. Like the celestial poles that are located directly above Earth's poles, the celestial equator lies directly above Earth's equator.

Annual motion

As Earth's rotation creates a daily cycle of objects rising and setting, Earth's motion as it orbits around the Sun also causes changes in the night sky over the course of a year. Because they are slower and a bit harder to perceive, detecting these changes in the position of objects in the night sky requires some record keeping. However, over several months of watching the night sky it readily becomes apparent that new constellations are indeed becoming visible in the east while once visible constellations are disappearing below the western horizon.

Stepping out at sunset each day and looking to the west, we notice that the constellations are steadily getting closer to the western horizon until they finally disappear altogether. Similarly, looking towards the east just before sunrise we notice new constellations rising above the horizon earlier and earlier each day. These are constellations which we have not seen since they disappeared below the western horizon half a year earlier. Over the course of a year, the constellations parade across the night sky. Since the apparent rotation of the celestial sphere repeats on a yearly cycle, we call this "annual motion".

The ecliptic

Earth's orbit around the Sun defines a plane. The intersection of this plane with the celestial sphere defines a great circle called the "ecliptic". Since the plane of Earth's orbit is centred on the Sun, from the perspective of someone standing on Earth, the Sun will always be found on the ecliptic. Put another way around, the path of the Sun in the sky "defines" the ecliptic. The name "ecliptic" comes from the fact that an eclipse of the Sun or the Moon can only occur when a new or a full Moon is crossing the ecliptic.

Now, if Earth's rotational axis were simply perpendicular to the orbital plane then the ecliptic and the celestial equator would be one and the same. In the real word, it is not so simple. Earth's rotation is not perpendicular – that is, at a 90° angle with respect to the orbital plane; instead it is "tipped" away from perpendicular by an angle of 23.4°. This results in the ecliptic also being inclined to the celestial equator by 23.4°.

Now, here is a peculiar property of a rotating spheroid like our Earth that makes it resemble a gyroscope. Spinning objects, like gyroscopes, tend to keep their axis of rotation fixed in space relative to the rest of the universe. If we move them up or down, left or right, their axis of rotation stays pointing in the same direction. The same holds true for Earth's rotational axis. As we have discussed earlier, the north celestial pole is a point in the sky, presently coinciding with the location of the star Polaris – that is, directly above Earth's north rotational pole. The daily motion of the stars in the northern celestial sphere always circles around Polaris. Even as Earth travels around the Sun, Earth's rotational axis remains fixed on Polaris. This means that at one point along the orbit, Earth's rotational axis is tipped towards the Sun, while half a year later, on the "other side" of its orbit, the axis is pointing away from the Sun. In the northern hemisphere, the rotational axis points towards the Sun during the summer and is tipped away from the Sun during the winter. When Earth's axis is pointed towards the Sun, the Sun appears above the celestial equator. Half a year later, when the axis points away from the Sun, the Sun appears below the celestial equator.

Solstices and equinoxes

For those of us who live far from the Equator, the tip of Earth's axis also noticeably affects the length of the day relative to the night as well as how "high" or elevated above the horizon the Sun appears at noontime. In the summer, when the rotational axis points towards the Sun, the noonday Sun climbs high in the sky, and the days are longer than the nights. The opposite is true in the winter, when the rotational axis points away from the Sun: the noonday Sun stays low in the sky, the days are now short, and the nights are long. The longest day of the year occurs when the noonday Sun reaches its highest elevation above the horizon, and the shortest day when the noon day Sun peaks at its lowest elevation. These two days are called the summer solstice and the winter solstice, respectively.

Between the two solstices are points where the ecliptic intersects the celestial equator. Both points are called an equinox since, on the day the Sun crosses the celestial equator, day and night are very close to being equal in length. After the Sun passes the winter solstice, the days begin to lengthen and the nights to shorten. The length of day and night become approximately equal at the vernal (Spring) equinox. The days continue to become longer at the expense of the night until they reach their maximum length at the summer solstice. As the days shorten again, day and night become virtually equal in length at the autumnal (Fall) equinox.

Planets and the zodiac

Early on, people noted that there were a handful of star-like objects seemingly dancing to their own tune. These stars appeared to meander about the sky, and so the Greeks called them *planētai* "wanderer".[25] The non-wandering stars, whose distances from Earth are so vast that they appear "fixed" to the celestial sphere, always move together as a group. The planets, though, wander about. Sometimes a planet's motion even runs counter to the general drift of the night sky.

We know that the planets are objects that orbit the Sun. Earth itself is one of eight planets (nine, if you count Pluto) that orbit our nearby star Sol. Mercury is the closest to the Sun, and then comes Venus. Mercury and Venus, because their orbits lie within Earth's orbit, are sometimes referred to as "inferior planets". We live on the third rock from the Sun. After Earth, there are the outer planets, which are sometimes referred to as the "superior planets" because their orbits are outside Earth's orbit. In order from Earth, these planets are Mars, Jupiter, Saturn, Uranus and Neptune (if Pluto still counted as a planet, it would be next).

People also noticed how, besides wandering around the sky, the planets were always to be found in a narrow band of the sky centred above and below the ecliptic. Recall that the ecliptic is the path the Sun appears to take across the celestial sphere over the course of a year. If these other orbits were all

[25] Arny, Thomas, *Explorations: An Introduction to Astronomy* (St Louis, MO: Mosby-Year Book, 1994) p. 27.

in the same plane as Earth's orbit, then the planets would all appear along the ecliptic. Of course, things cannot be that simple. Each planet's orbital plane is actually tilted, albeit very slightly, relative to Earth's orbital plane. This results in the planets' positions in the sky alternating above and then below the ecliptic, defining a very narrow band which is called the zodiac. The Sun, the Moon and all the planets are always found somewhere within the band of the zodiac.

Going retrograde

A planet's apparent motion in the sky is due to Earth's orbital motion around the Sun as well as each planet's own orbital motion. From night to night, the stars appear to drift from east to west, but the planets normally drift in the opposite direction, from west to east as they orbit the Sun. Occasionally, a planet appears to stop its eastward drift, pause, then drift westward, in the same direction that the sky moves, pause again, and then return to its original easterly course. When a planet changes direction and backtracks towards the west, this is called retrograde motion.

If we were to look down on the solar system from the north, we would observe the planets orbiting the Sun counterclockwise. The inner planets will be orbiting faster than the outer planets. Now, as Earth catches up with a slower outer planet, such as Mars, Jupiter and Saturn, and passes it, the planet's motion appears to momentarily pause against the backdrop of the more distant stars. It then reverses direction for a short time, retracing its path, before stopping a second time and then resuming its normal easterly course. When a planet changes direction and drifts west, we call this "retrograde motion." When the planet pauses as it changes direction, we call this a "station".

Astronomy, astrometry and astrology

At least 4,000 years ago, our ancestors were well aware of the sky's annual motion and were keeping records of what they observed. One reason for recording their observations was to keep track of the time of year without necessarily counting the days. For those who lived far "above" or "below" the equator, determining the time of year did not seem so difficult. If they looked outside and saw ice and snow, it was winter. If the temperature and humidity were uncomfortable, it was surely summer. Spring happened in between winter and summer, and autumn would follow summer and lead back into winter. However, for people living closer to the Equator, such as the Egyptians and Mesopotamians, the yearly fluctuation in temperature could be much more subtle.

For the Egyptians and Mesopotamians, reliably knowing when to plant their crops, move their flocks to different pastures or even venture out on ocean voyages required accurate knowledge of the time of year. The annual parade of constellations across the night sky provided them with a yearly calendar. When Leo was in the evening sky, it was getting close to planting time. It was surely summertime when Scorpius appeared. Likewise, when Orion the Hunter was high in the night sky, it was winter. In addition to constellations, individual stars could also be used. For instance, the Egyptians knew that when the bright star, Sirius, in the constellation Canis Major, first became visible in the east just before sunrise (referred to as a "heliacal rising"), the Nile would soon flood. The heliacal rising of Sirius would signal that the time had come to plant crops.

So, "reading" the sky as a clock or calendar became an important human activity. Celestial objects rising in the east and setting in the west defined a day. The Moon repeatedly waxes (increases in size) until it is a full disc and then wanes (decreases in size) until it vanishes and then reappears again to repeat the cycle. The period of time it takes the Moon to wax and wane is about twenty-nine days – a useful unit of time for organising activities, which would eventually become the span of time we call a month.[26] Finally, each day the Sun appears to "drift" a little towards the east relative to the background stars. Over the course of about 365 days the Sun circumnavigates the celestial sphere, returning to the same spot from where it began its journey. This unit of time became a year.

---

[26] The twenty-nine days would be a synodic lunar month, the time necessary to go through a full wax and wane cycle as viewed from Earth. There are all sorts of other lunar-based months. The Gregorian Calendar months we have today are of similar length to the lunar months but are driven more by a desire to divide a year into twelve somewhat equal sections.

As the scribes in ancient Babylon watched and recorded the movement of the celestial sphere and noted the motions of the Sun, the Moon and the planets relative to the distant stars, they were doing then what we would today recognise as astronomy: the study of objects and matter outside Earth's atmosphere. To be more precise, they were practising a branch of astronomy called astrometry, which deals with measurements such as the positions and movements of celestial bodies.

Watching the sky became extremely helpful in planning activities such as when to plant crops or to move the flocks to new pastures. Those who sailed on ships knew that there were seasons of the year when storms were frequent and other seasons when the weather was quieter. For sailors, knowing what time of the year they were in might be a matter of life or death. In an age before smartphones that tell you the year, the month, the day of the month and the time of day, watching the sky was your clock and calendar. Astrometry was an important human activity. But you do not have to watch the daily, monthly and yearly cycles of the sky for very long before you begin to associate the patterns of the sky with the patterns of life on Earth.

As our ancestors realised that the annual motion of the Sun, Moon, planets and stars in the night sky was predictable and could be relied upon to tell them to, let us say, start planting, a connection was then made between the heavens and the lives of humans on Earth. Putting it bluntly, people began to order their lives according to celestial events. As time passed and people found ever more useful ways the signs in the sky could help them organise their activities on Earth, the linkage between the stars and their fortunes became even stronger. I would argue that it was not difficult to go from looking to the stars to help plan your day to relying on them to plan your life.

We have now arrived at a chicken and egg moment: did astronomers start doing a little divining "on the side" to make a living for themselves or did astrologers start developing mathematical astronomy to be able to predict future celestial phenomena and give their clients more lead time for what the fates had in store? I am tempted to answer yes and yes. Instead, I will remain agnostic on this question. What is important to note is this: from antiquity up till the time of Johannes Kepler (AD 1571-1630) – an astronomer whose laws of orbital motion aid today's spaceships to navigate the solar system – terms such as "astrology" and "astronomy" were interchangeable. Astronomers and astrologers were the same individuals. As already mentioned, old Johannes himself is known to have cast a few horoscopes. So, for our purpose, whether we refer to someone as an astronomer or an astrologer will depend on whether they are focusing on making observations of material objects (astronomy) or divining from these observations omens and portents for their patrons (astrology).

Philosophy: Rational Reflection

It was in Ionia that the new Greek civilization arose: Ionia, in whom the old Aegean blood and spirit most survived, taught the new Greece, gave her coined money and letters, art and poesy, and her shipmen, forcing the Phoenicians from before them, carried her new culture to what were then deemed the ends of the earth."

H.R. Hall, *The Ancient History of the Near East*

Wise men in ancient Babylon and Egypt were carefully observing the world around them. We know this because they recorded their observations and their thoughts, and somehow these writings have managed to survive for thousands of years until today. Our earliest Babylonian star catalogues date from around 1200 BC. Using records like these, we can trace mathematics and geometry back to at least this time. Our wise men were developing these topics partly to help with practical problems such as determining boundaries between fields after the Nile's annual flood, but also to help in the forecasting of omens, for example by anticipating the heliacal rising of a star or planet. Their astronomy was more analogous to "lore" than to the detailed understanding of how, say, gravity dictates the motions of bodies observed in the celestial sphere. The mathematics, astronomy and geometry they developed were more akin to "rules of thumb" and "first order approximations" than precise formal descriptions of the phenomena observed.

Our modern understanding of scientific mathematics and astronomy derives from the Greeks. While it is likely that Greeks of the coastal Adriatic gained knowledge from the wise men of the Nile and the Euphrates, the Greeks took their inquiries into a new direction that we can recognise today as science and philosophy. As Fr Frederick Copleston SJ observes in his *A History of Philosophy*, the beginnings of mathematics and astronomy, "Science and Thought, as distinct from mere practical calculation and astrological lore, were the result of the Greek genius and were due neither to the Egyptians nor to the Babylonians".[27]

From the rivers of Babylon to the sun-drenched Aegean shores

Thales of Miletus is credited as the first philosopher. He was also a mathematician and astronomer who, it is said, predicted an eclipse of the Sun mentioned by the Greek historian and geographer Herodotus as happening at the end of a war between the Lydians and the Medes. Calculations tell

[27] Copleston, Frederick, *A History of Philosophy*, Vol. 1 (New York: Doubleday, 1993) p. 16.

us that there was an eclipse of the Sun visible in Asia Minor in 585 BC. If this is the eclipse mentioned by Herodotus, it would appear that Thales lived during the early part of the sixth century BC. It is thought that he died just before the fall of Sardis, the decisive battle that led to Cyrus II and the Persians ruling over coastal Asia Minor, including Ionia. As we have already noted, the Persians "bumping" into the Greeks set off a chain of events that ultimately led to a Hellenised Jew of Edomite lineage being declared King of the Jews and placed on his throne in Jerusalem by an act of the Roman Senate. Still, that is five hundred or so years in the future, and so let us return to the beginnings of Hellenic culture.

Reality is One – Heraclitus

At the end of the sixth century BC, in the city of Ephesus on the Ionian coast, we encounter Heraclitus. He is known for his pithy sayings – some of which he may have actually said – such as "Asses prefer straw to gold" and "Man's character is his fate." Heraclitus had little use for the mystery religions or cults of his time, and his view of God was pantheistic – that is, that everything constitutes a unity, and the unity is divine. According to Frederick Copleston SJ, Heraclitus's original contribution to philosophy is "in the conception of unity in diversity, difference in unity".[28]

---

[28] *Ibid.*, p. 40.

Copleston continues, "That Reality is One for Heraclitus is shown clearly enough by his saying: 'It is wise to hearken, not to me, but to my Word, and to confess that all things are one.'"[29] Whereas some of his predecessors and critics saw difference and diversity as a defect that mars the beauty of the One, or the whole, Heraclitus sees diversity and difference as essential to the One. As Copleston points out, "The philosophy of Heraclitus corresponds much more to the idea of concrete universal, the One existing in the many. Identity in Difference."[30]

So, what is the "One in many", and who or what is this "One"? Well, before we delve deeper into these questions, we should first investigate what it means to label Heraclitus a pantheist.

Pantheism is a belief that God *is* everything; or, put the other way around, God is not in any way distinct from the universe. The word pantheism is built upon two Greek words: *Pan*, meaning "all", and *Theos*, which means "God". In this system of thought, the deity is identified as being one and the same with the universe and all its phenomena. Although Heraclitus sees the One as wise and universal Reason that animates all things, he does not see God as a personal being independent of the material world and its phenomena. For Heraclitus, and for pantheists in general,

[29] *Ibid.*, p. 40.

[30] *Ibid.*

God is an impersonal all-encompassing force that animates but is not independent of the material universe.

Now, getting back to what Heraclitus considers the "One in many", we next need to speak about the ancient idea of Elements. The classical Elements that were considered the fundamental constituents of all reality are Earth, Water, Air and Fire. Predecessors of Heraclitus at one time or another suggested that one or another of these elements was the fundamental essence of the One. Thales proclaimed that water was the primary stuff of all things.[31] Anaximander, thought to be younger than Thales, felt that none of the elements were primary, but instead left the primary element as indeterminate. Anaximenes, thought to be younger than Anaximander, came down on the side of Air being the primary substance of the One.

Heraclitus, and the Stoics who would rely heavily on his thoughts, came down on the side of Fire being the essence of all things. We might be tempted to think that Heraclitus merely picked the next primal Element on the list to differentiate himself and give his system of thought some novelty in relation to his predecessors. This might be a tempting idea, but on further reflection we can see that Heraclitus might have been more discerning than his predecessors. For one thing, it is our common experience that most things can be made to burn. Fire, in fact, feeds on

[31] *Ibid.*, p. 22.

lots of different types of matter, appearing to change them into itself. Take away the fuel and the fire will cease to exist. Fire requires "strife" and "tension" for its existence, as noted by Copleston.[32] Sense experience tells us that Fire lives by feeding on, consuming and transforming things into itself. Fire is both want and abundance. It is all things in a constant state of transformation from one Element to another. To explain the apparent stability of the world, constantly in flux, is explained by a detailed balance between Fire consuming and giving back from itself.

As mentioned earlier, Heraclitus's view of God was pantheistic – that is, all is divine. He also saw God as wise and the universal Reason (*Logos*) animating all things. God brings harmony to all things and directs all things according to universal law. When we say "all things" we mean people, too. Heraclitus taught that man's ability to reason is but a "moment" in the universal Reason. Man's reason was a small portion of the universal Reason. Therefore, understanding that all things are in unity and that universal law controls all things, man can strive to understand and come to terms with the viewpoint of Reason and so live by its all-comprehensive, all ordering *Logos* (Word) or law.

This will also require the followers of Heraclitus such as the Stoics to "come to terms" with the problem of evil; notably, if everything is God and God is wise and good,

---

32 *Ibid.*, p. 41.

then should not everything be perfect and good? However, human experience tells us that there are things about the world that are not good. But, then again, as a part of the universal Reason, how is it that we even come to the idea that things are not always as they should be?

A related question may also be raised. If humans are all part of God, then our actions must all be the actions of God, so how can universal Reason require of us any particular duty? Here we can see how the belief in a pantheistic God such as Reason or *Logos* seems to imply that humans are not free agents but merely components of a collective. Words like "duty" and "ought" lose their power to compel us if our actions are all predetermined.

Cyclical and linear time

When we start speaking of determinism, we should also think about time, and how Heraclitus, the Stoics and, for that matter, most ancients viewed time. It appears that, historically, most humans have regarded time as cyclical. The cycling of the seasons – summer, autumn, winter, spring and back to summer again – form a yearly cycle.

Closely related to the cycling of the seasons is the agricultural cycle of planting, growing, reaping and planting again. As we have seen, there are the cyclical patterns in the heavens, such as the daily rising and setting of the Sun, the monthly waxing and waning of the Moon, the yearly voyage of the Sun around the band of the zodiac, and the motions of the planets, with cycles on the order of years to tens of years. The experience of these cycles, as well as many more that I am sure the reader could identify, naturally leads to a cyclical or circular view of time. This was especially true for people living close to nature and for whom the agricultural and seasonal cycles directly prescribed their activities at any given time, which would often repeat without much variation for the entirety of their lives.

Cyclical time was an experience not just of the individual, but of their cultures, too. Many cultures developed the concept of a cyclical birth, life, death and rebirth of the universe. These universal cycles were much, much longer than, let us say, the nearly thirty-year cycle of Saturn's orbit about the Sun. For instance, the Hindus of the Indus region had a cycle called "one life of the Brahma", a period that spanned the birth, life, death and rebirth of the universe, 311 trillion (311,000,000,000,000) years.[33] This is a respectably large number even for a modern astronomer. According to the Han period mathematical text known as *Chou Pei Suan Ching*, the Chinese had the Grand Period of 31,920 years when every process returned to its original form or condition.[34] Finally, the Greeks had the Great Cycle, said to have a length of 10,800 years.[35]

---

[33] Jake, Stanley L., *Science & Creation* (Edinburgh: Scottish University Press, 1986) p. 3.

[34] *Ibid.*, p. 33.

[35] *Ibid.*, p. 107.

The Stoics taught that the Great Cycle ended with a universal conflagration followed by a rejuvenation of the world. They also taught that each reincarnation of the world would repeat, down to the last detail, everything that occurred in its previous incarnation. This included the actions of every person. Each individual would repeat every action of their lives over again in each cycle.

It is our friend Heraclitus who is credited with determining the Great Cycle's length of 10,800 years. Even if this is true, it is doubtful that he believed the Stoic's doctrine of universal conflagration. The idea of a cyclical repetition of the world being the same in every detail seems far removed from the man to whom the phrase "for it is impossible to step twice into the same river" is famously attributed.[36] Yet Heraclitus embraced the Great Cycle.

A tendency to conceive of time as cyclical is easily understood. Our everyday experience of seasons supports such a notion. That cultures can also conceive time on the cosmological scale as being cyclical is clearly true, given the historical examples we have just enumerated. It is curious, then, that Judeo-Christian culture does not see time as fundamentally cyclical but, instead, linear. For Jews and Christians, time has a definite beginning and is moving to a definite end. Even the modern secular culture that identifies

[36] Frag. 91, see Kirk, G.S., *Heraclitus, The Cosmic Fragments* (Cambridge: The University Press, 1954) p. 381.

as progressive shares the linear view of time bequeathed to them by their Judeo-Christian forebears. "Little by little, day by day, getting better and better in every way" would imply a linear view of time, since cyclical time would eventually mean that you progressed back to where you started.

The Judeo-Christian view of time might very well be unique. Be that as it may, I would suggest that a culture with a linear view of time finds it harder to believe in divination and astrology than a culture with a cyclical idea of time. For those who have a cyclical view of time, with human events recurring endlessly over and over again, the ideas of divination and astrology may seem much more plausible than they do to us. If cyclical time is also coupled with the concept of a deterministic universe, where there is no room for free will, then divination and astrology may seem not only plausible, but probable.

Stoicism

Stoicism traces back to Zeno of Citium (not to be confused with Zeno of Elea, of paradox fame) who lived sometime between 334 and 262 BC He was born in Citium (present day Larnaka) on the southeastern shores of the island of Cyprus.[37] Zeno appears to have spent the early part of his life in the commercial sector and may have been wealthy.

---

[37] Most of what we know of Zeno comes from a biography and anecdotes recorded by Diogenes Laërtius, who lived in the third century AD. Most of what we think we know of ancient Greek philosophers comes from his work, *Lives and Opinions of Eminent Philosophers.*

While visiting Athens, he was introduced to the thought of Socrates. As the story is told, Zeno read of Socrates and was so enamoured with what he read that he asked the bookseller where he could find such a man that day in Athens. Just then, Crates of Thebes happened to be passing by, and the bookseller pointed to him. Zeno began his training in rational reflection – what we call philosophy – with Crates. He also studied with other leading thinkers of the day in Athens.

When Zeno himself began teaching, he set up shop in the colonnade in the Agora of Athens known as the Stoa Poikile or Painted Porch. The Agora, a market or gathering place, was a centre of public life in the city. Not only was it a place of commercial activity, but it was also the location for political, artistic, athletic and spiritual life. Zeno's students and followers would eventually be called Stoics, and his system of thought would become known as Stoicism.

Two followers of Zeno helped establish Zeno's philosophy as an enduring school of thought. Cleanthes of Assos (c.330-c.230 BC) succeeded Zeno. Cleanthes was followed by Chrysippus of Soli, who is said to be the second founder of Stoicism by virtue of his systemisation of Stoic doctrines.[38] These founders of Stoicism managed to put together a system that embraced "all things divine and

---

[38] Copleston, Frederick, *A History of Philosophy*, Vol. 1 (New York: Doubleday, 1993) p. 385.

human",[39] which would make it attractive to many people looking for a "way" to order their lives. Notably, Stoicism was very attractive to the Roman elites at the start of the first millennium, such people as Emperor Marcus Aurelius.

Maxwell Staniforth concisely lays out the key ideas of Stoicism in the introduction to his translations of Marcus Aurelius's *Meditations*. According to Staniforth:

> The three keywords of Zeno's creed were Materialism, Monism, and Mutation. That is to say, he held that everything in the universe – even time, even thought – has some kind of bodily substance (materialism); that everything can ultimately be referred to a single unifying principle (monism); and that everything is perpetually in process of changing and becoming something different from what it was before (mutation).[40]

Stoics did not accept the way that Plato and his student Aristotle divided the world between perceptible objects, such as a "horse", and transcendent universals such as "horse-ness". This would seem to be consistent with the Stoics' idea that all things, even thought, are material. For them only individual, material objects exist, and our knowledge, therefore, is only of particular objects. They would say that they have seen a "horse" but have never seen "horse-ness".

---

[39] Marcus Aurelius, *Meditations,* tr. Maxwell Staniforth (London: Penguin Books, 1964) p. 9.

[40] *Ibid.*, 9-10.

The Stoic's cosmology (the world as a totality of space, time and phenomena) was based on concepts developed by our old friend, Heraclitus. Like Heraclitus, the Stoics were materialists who subscribed to the theory that the element Fire was the substance from which all things, including time and the soul, were derived. Stoics were pantheists. They believed that God and the world were one and the same. They also followed Heraclitus's conception of the *Logos,* the Word, as the ordering principle of the world. They believed that nature is rational and that the universe is guided by the law of reason.

This Reason or God, immanent in matter and having no existence outside of the material, is the animating and guiding force in the universe. The Stoics believed that God arranged the universe for the good of man. They also saw man as the "highest phenomena of nature", and since man is but a constituent part of the whole universe and possesses consciousness, then it only seems right that the world possesses consciousness, too. God, then, is the Consciousness of the world.[41]

Up to this point, the Stoics and Heraclitus follow each other, but we now come to a concept that is important to our understanding of the role astrology and divination played in the lives of first-century peoples, with the notable exception

---

[41] Copleston, Frederick, *A History of Philosophy*, Vol. 1 (New York: Doubleday, 1993) p. 388.

of the Jews. The doctrine the Stoics held to, but which Heraclitus is unlikely to have professed, is that of universal conflagration. As Copleston explains:

The Stoics, however, certainly added this doctrine of the *ekpyrosis*, according to which God forms the world and then takes it back into Himself through a universal conflagration, so that there is an unending series of world-construction and world-deconstructions. Moreover, each new world resembles its predecessor in all particulars, every individual man, for example, occurring in each successive world and performing the identical action that he performed in his previous existence.[42]

It is easy to understand, given such a cyclical view of time and existence, how the Stoics did not believe in human freedom. Free will is a concept incompatible with a thoroughly deterministic universe. The Stoics referred to this determinism as Fate. Fate and Providence, which orders all things for the best, are two aspects of God. The Stoics would point out that, although the world is deterministic, and a man's life is not free but preordained, nevertheless, an individual can choose to assent to the actions Fate ordains and welcome them as "God's will". In a word, a Stoic strives to be "fatalistic".

For the Stoics, this deterministic and completely interconnected universe allowed for divination, portents

---

42 *Ibid.*, p. 389.

and oracles. As Copleston says, "Because of the universal harmony of the cosmos and the reign of Fate the future can be divined in the present: moreover, the Providence of God would not have withheld from men the means of divining future events."[43] As we will see, with the notable exception of the Jews, many people at the turn of the first millennium would make use of varied forms of divination, in particular astrology, to try and learn what the Fates had in store.

Sound familiar?

In our discussion of Heraclitus and Stoicism, we Christians are hearing familiar words such as Logos, which cannot but make us think of the Word of God who will be incarnate and become man. We hear of small flames breaking off from the "Divine Fire of Reason" and animating humans and cannot help but imagine the locked room at Pentecost with tongues of flame descending on the apostles. But we must not confuse the "Divine Fire" of Heraclitus with any notions of a personal God. As previously mentioned, Reason is a pantheistic God. For Heraclitus, God is the immanent, ordering principle of all things. Heraclitus's goal was, in part, to help his disciples to come to terms with and resign themselves to what Nature was going to do with or without their assent. Unlike the Christian, the Stoic was required to abandon hope.

---

[43] *Ibid.*, p. 424.

Philosophy as religion

Unlike the Hebrew people, the Greeks and Romans of classical times turned to philosophy for guidance in answering such questions as how to live a good life, the difference between good and evil and the nature of our responsibilities to our neighbour. Today, we believe that these and other big questions – such as the origins of the material world and what the ultimate end to human life is – fall squarely in the domain of religion, whereas for many Gentiles at the beginning of the first millennium it was the philosopher they turned to and not the priest. In the time of Imperial Rome, the priest was a functionary of the government whose job was to make the prescribed rituals at their proper times in order to ensure the gods' protection for the state, or, at the very least, to assuage their displeasure. Morality, ethics and the meaning of life were not the concern of Roman religion or its priests.[44]

Questions about life and its meaning were the concern of the philosopher. As today there are many religions vying for followers, so was the case with competing philosophies at the start of the first millennium. They each had their own take on what the answers were to life's big questions. In all cases, though, it was the philosopher who fulfilled the role that rabbis, priests and ministers do today.

---

[44] Marcus Aurelius, *Meditations*, tr. Maxwell Staniforth (London: Penguin Books, 1964) p. 8.

The philosophy of Stoicism, though, had a religious air and became very influential.[45] With its practical and moral principles, along with its teachings on man's affinity with God and obligations for the well-being of his neighbour, Stoicism was an attractive philosophy to Romans of the early Roman Empire.[46] In his two-volume history of European morals, William Lecky tells how, for the Imperial Romans, "Stoicism, taught by Panætius of Rhodes, and soon after by the Syrian Posidonius, became the true religion of the educated classes. It furnished the principles of virtue, coloured the noblest literature of the time, and guided all the developments of moral enthusiasm."

---

[45] *Ibid.*, p. 9.

[46] Copleston, Frederick, *A History of Philosophy*, Vol. 1 (New York: Doubleday, 1993) p. 428.

Astrology in the First Century BC

By the first century BC, Babylonian astrology, which relied on observing celestial bodies in the sky and then interpreting their meaning, had become completely passé and was replaced by a thoroughly Greek form of astrology that relied on calculating the positions of the planets and stars. The ability to calculate the positions of the stars and planets and their times of rising and setting had this major implication: astrologers could now cast horoscopes for events that happened in the past or that would happen in the future or that occurred any place on the Earth's surface.

The philosophy of Stoicism formed the basis for Greek astrology. For Stoics, a deterministic and completely interconnected universe allowed for divination, portents and oracles. As we have seen in the previous chapter, the philosophy was extremely popular at the beginning of the first millennium, particularly amongst the Romans, and especially among the Roman elite. Stoicism taught that the universe was conscious, and it controlled all things for the good of man through divine Reason. They called the ordering of all things for the best "Providence". The ordering of all things for the best also meant that the world was deterministic. They called this determinism "Fate".

Fate is an aspect of the Universal Reason, that impersonal and all-encompassing force that Stoics believed animates, but is not independent of, the material world. For Stoics, the essence of the material world is Fire, consuming and giving back from itself. Man is rational because he shares a small portion of this divine Fire, this Reason. Since Reason orders the world for the good of man, the Stoics believed that the Providence of the Universal Reason would not withhold from men the means of divining future events.

The Stoics also believed time to be cyclical. Their universe cycled through an endless series of world constructions and world deconstructions. Since Reason guides all things, the universe evolves in the best possible way, and so every cycle is thought to be identical to the previous cycle.

Providence, the Stoics believed, would not withhold from men the means of divining future events. Divination could reveal the will of Fate as it guided all events, and divination was the stock-in-trade of the Magi. For if a Magus could predict what Fate had in store for you, then you could prepare yourself to meet it and make the best of it. Forewarned is forearmed.

The zodiac

Ancient stargazers noted that the paths of the Sun, Moon and planets are restricted to a band of the sky. The path traced out by the Sun against the backdrop of the celestial sphere is the ecliptic. The planets and the Moon are always in a narrow band of sky centred on the ecliptic. This band they called the zodiac. The zodiac is divided into twelve sections, called "signs", each one spanning thirty degrees. These are the astrological signs most of us are familiar with: Leo (♌), Virgo (♍), Libra (♎) and so on.

Judaea and Aries

Now astrologers of the time believed that each city, nation and peoples were under the control of a particular sign. The three major sources we have for understanding first century astrology, Claudius Ptolemy of Alexandria, Vettius Valens of Antioch and Marcus Manilius of Rome, all agree that Herod the Great's kingdom was ruled by the sign Aries, the ram. If there were a portent involving Judaea, it would appear in the sign of Aries.

This is where the mysterious roman coins enter our story.

Power: follow the money

During most of the period of the Roman Empire, cities and client kingdoms minted their own coinage. Typically, the obverse (the "heads" side) would picture the emperor or some local deity. The reverse side was used to promote various local political or religious themes. These practices endure to our very day, as beautifully demonstrated by the "50 States Series" of quarters struck by the U.S. mint. On the obverse side we have a portrait of the American Caesar,

George Washington. On the reverse side, we find symbols promoting various local political and cultural themes that identify a particular state. Massachusetts is identified by an outline of the state and a representation of Daniel Chester French's famous minuteman statue. In the U.K., the fifty-pence coin has been regularly used since it was introduced in 1969 to commemorate important national events. The obverse of any new design always bears a profile of the current monarch, but unique reverse designs are sometimes minted, for events such as the fiftieth anniversary of the D-Day Landings in 1994 or that of the National Health Service in 1998.

In Roman times, there are instances of cities or countries identifying themselves by using the sign of the zodiac that ruled over them on their coins. For example, the Syrian city of Antioch used the astrological sign Aries, the ram, on their coins for two hundred years. Interestingly, the sign first appeared on their coins in AD 6, the year that Rome finally gave up any pretence of there being a Jewish king and annexed Judaea to Syria.

After Herod the Great's death in 4 BC, his kingdom was divided amongst his sons, but they were never given the title "King". His son Herod Archelaus was given the regions of Judaea, Samaria and Idumea. Eventually, Rome became so exasperated with the constant complaints about his despotic rule that it deposed Archelaus and sent in Quirinius, the governor of Syria, to bring Judaea under direct Roman rule.

Bronze coins depicting a leaping ram looking back over its shoulder at a star first begin to appear on coins minted in Antioch c. AD 5-11. This example comes from the Molnar collection, PRC-1 4265.

Note that this is the same Quirinius named in Luke's Gospel as the governor of Syria at the time of Jesus's birth.

So, Aries the ram showed up on coins struck in Antioch, the capital of the Roman province of Syria, close to, if not during, the time of Quirinius's governorship. These bronze coins, referred to as Tríchalkons, are of low denomination and so presumably were in wide circulation in the Syrian province and the regions it administered. One side of the coin portrays a headshot of Zeus. The other side, though, depicts a leaping ram looking back at a star. So, why would the governor of Syria place this particular astrological symbol (Aries the Ram) looking back at a star on the coin?

Part of the answer has to do with good old-fashioned power politics.

As we outlined above, the Hellenistic kingdom of Syria had once ruled Judaea. The Jews managed to overthrow their Syrian overlords and to maintain their independence for a century or so. All pretence of an independent Judaea vanished, however, when Rome deposed Herod Archelaus and turned the administration of Judaea over to Syria. The governor of Syria then minted new coins prominently displaying a leaping ram, Aries, the astrological sign for Judaea. The point should be clear. The coins are saying "We're back!"

Now, setting Syrian national pride aside for a moment, the coin is tangible evidence that Aries the ram was the sign for Judaea. Another version of the coin appears in AD 13-14, minted under governor Silanus. Again, depicted on the coin is a leaping ram looking back over its shoulder at a star. These so-called Aries coins will be minted for the next two hundred years. So the question is, why the star?

Well, whatever the star's meaning turns out to be, always remember this key point:

> "Roman provincial coinage served as a primary medium for disseminating propaganda supporting the goals of Rome." (Dr Molnar)[47]

As a collector of ancient coins Dr Molnar knew this, and when his research turned up that Aries was the astrological

[47] Molnar, Michael R., *The Star of Bethlehem: The Legacy of the Magi* (New Brunswick, NJ: Rutgers University Press, 1999) p. 48.

sign for Judaea, he asked himself whether Rome could perhaps have been sensitive to rumours amongst astrologers of a regal portent in Aries, which the locals believed heralded a Messiah? If so, could the coins be an attempt by Rome to subvert that apparition by saying that, yes, there was a regal portent in Aries, but it signified Caesar's abolishing Judaea and not a Jewish Messiah? Dr Molnar realised that if these coins were linked to the star of Matthew's Gospel, then they were tangible evidence that Matthew's star was likely to have been a historical event and not a myth.

Molnar now knew where to look for the star – the zodiac sign of Aries – but he still needed to know when Jesus was born and what, exactly, would the Magi have considered an iron-clad portent for the birth of a king.

Playing Astrological Detective

So, Jesus was born in the year AD 1, right? Well...it's a bit more complicated than that.

Counting the years

Our modern calendar, which simply numbers the years since the Incarnation, was first proposed by a monk named Dionysius Exiguus, in AD 525. Pope John I charged Dionysius to extend the table of dates for Easter for another ninety-five years. In the course of extending the tables, the monk decided to switch the counting of years from the Diocletian Era, also called the "Era of the Martyrs", to a new system he invented that numbered the years from "the incarnation of our Lord Jesus Christ". Possibly because Roman numerals did not include a "zero", the year of the Incarnation was intended to be the first day of AD 1 rather than AD 0. The years before the Incarnation are numbered backwards beginning with year 1 BC.

Dionysius explained that he created this new Christian calendar so as not to perpetuate the memory of an emperor who persecuted Christians. Now, just how he determined that 525 years had passed between the Incarnation and Pope John's commissioning him to extend the table of Easter dates is not known. To further confuse things, we are not even certain what he considered to be the Incarnation? Since he told us he was numbering the years from the Incarnation, then he most likely meant the Annunciation, but perhaps he was thinking of the Nativity, nine months later. A discrepancy of nine months or so may not seem to be much of a concern, but it is still three-quarters of a year, and there are ancient events whose dates we do know precisely. For instance, the date of an event that happens to coincide with a lunar or solar eclipse can be precisely determined. Recall that Herodotus told of how a conflict between the Lydians and the Medes quickly came to a halt because of a solar eclipse. If Herodotus's report of the so-called Battle of the Eclipse can be taken at face value, then the battle ended late in the day of 28 May 585 BC.

Another example of an eclipse helping to determine the date of a historical event is the death of Herod the Great. In his book, *Antiquities of the Jews,* Josephus tells how Herod died shortly after a lunar eclipse and just before the Passover. Though several lunar eclipses occurred during this period, the eclipse of 13th March 4 BC seems to be the best fit. I think the reader can now begin to see the problem: in Dionysius's system of numbering the years from the Incarnation, Herod has apparently been dead for four years prior to the Incarnation in AD 1, yet both Matthew and Luke assure us that Jesus was born during the reign of Herod the Great. It would appear, therefore, that Dionysius's numbering system is a bit off.

Plausible range of dates for the Nativity

To try and ascertain a plausible range of years for Jesus's birth, we will closely follow the investigation carried out by Dr Molnar.[48] We will work with his list of datable historic events and personages thought to be contemporaneous or in some way linked to the time of Jesus's birth. We will add to Dr Molnar's list another datable event, the finding of the child Jesus in the Temple. We will then see where all these events overlap. The span of years that includes most or all of what we know should allow us to determine in our summing up a likely range of years that would encompass the date of Jesus's birth.

Reign of King Herod the Great

Both the Gospels of Matthew and Luke tell us that Jesus was born during the reign of King Herod. Matthew states:

Now when Jesus was born in Bethlehem of Judaea in the days of Herod the king, behold, Wise Men from the East came to Jerusalem, saying, "Where is he who has been born king of the Jews? For we have seen his star in the East, and have come to worship him." (*Mt* 2:1-2)

---

[48] Molnar, Michael R., *The Star of Bethlehem: The Legacy of the Magi*, (New Brunswick, NJ: Rutgers University Press, 1999) pp. 57-63.

This would seem to be an unambiguous statement that Herod was alive at the time of Jesus's birth.

Herod's reign was a long one, beginning with the Roman Senate declaring him King of the Jews in 40 BC and his final defeat of the last Hasmonean ruler, Antigonus, three years later in 37 BC. From the Judaean perspective, then, we will say that Herod's reign begins when he attains practical authority in 37 BC and ends with his death in 4 BC. The birth of Jesus could have occurred anytime during the reign of King Herod and so 37-4 BC is our first "benchmark" or range of years for the birth of Jesus.

Public lives of John the Baptist and Jesus Christ

Luke provides us with another benchmark when he tells how John the Baptist began his public ministry in the fifteenth year of the reign of Tiberius Caesar, which is considered to be around AD 28-29 (*Lk* 3:1). Luke also tells us that Jesus began his public ministry when he was "about thirty years of age" (*Lk* 3:23). If we take "about thirty" to mean plus or minus three years, then we can say that Jesus began his ministry between the ages of twenty-seven and thirty-three years old. Combining this range of ages with the range for the fifteenth year of Tiberius's reign to get a maximum range of possible birth years for Jesus, we arrive at a benchmark of 6 BC-AD 2.

Conception of John the Baptist

We also learn from Luke's Gospel that Jesus was born six months after John. So, what do we know about John's birth?

During the reign of King Herod, a priest named Zachariah and his wife Elizabeth, both well along in years and childless, conceived a son who would be known as John the Baptist (*Lk* 1:5-25). We are also told that in the sixth month of this surprise pregnancy, the Angel Gabriel announced to Elizabeth's kinswoman, Mary, that she was to conceive and bear a son whom she should name "Jesus" (*Lk* 1:31). The Angel Gabriel went on to inform Mary that her older relative, Elizabeth, who was thought to be barren, had also conceived and was now in her sixth month of pregnancy (*Lk* 1:36).

From these two data points – the conception of John the Baptist and the ensuing conception of Jesus six months later – we can deduce that Jesus was born approximately fifteen months after the conception of John the Baptist. Since there could be a little uncertainty in these numbers (a baby's arrival can sometimes be early and other times be late), let us say that Jesus was born fifteen to sixteen months after the conception of John the Baptist. If John's conception occurred at the same time as the death of Herod in April of 4 BC, then the latest Jesus could have been born is sixteen months later, in August of 3 BC. We can use this datum to place a limit on how late Jesus could have been born on the range we obtained from considering Jesus as being "around thirty" in the fifteenth year of Tiberius. The modified benchmark we will use is 6 BC-3 BC.

Tax census during Quirinius's governorship of Syria

Now we come to a problem. As we have stated above, the Gospel of Matthew states explicitly that Jesus was born during the reign of King Herod the Great. The first chapter of Luke's Gospel clearly implies that Jesus was born during Herod's reign or within sixteen months after his death. But then, in chapter 2, Luke goes on to report:

> In those days a decree went out from Caesar Augustus that all the world should be enrolled. This was the first enrolment, when Quirinius was governor of Syria. And all went to be enrolled, each to his own city. And Joseph also went up from Galilee, from the city of Nazareth to Judaea, to the city of David, which is called Bethlehem, because he was of the house and lineage of David, to be enrolled with Mary his betrothed, who was with child. And while they were there, the time came for her to be delivered. And she gave birth to her first-born son and wrapped him in swaddling clothes, and laid him in a manger, because there was no place for them in the inn. (*Lk* 2:1-7)

The problem is this: Sulpicius Quirinius became governor of the Roman province of Syria with authority over Judaea beginning in AD 6, nearly a decade after the death of King Herod, so how can we reconcile Luke's chronology?

Attempts to explain Luke's timeline often centre on the census or enrolment he mentions in Luke 2:1-2 and whether or not the evangelist was confused as to which of the many enrolments that had occurred by the time of Jesus's birth was the reason for Joseph taking Mary to his family's hometown of Bethlehem.

Quirinius was indeed assigned by the emperor the task of conducting a census when he was sent to Syria in AD 6. Technically, Quirinius was appointed not as governor but as legate; an official by the name of Coponius was the governor (AD 6-9).[49] In any case, Quirinius as legate was a high-ranking military official in charge of the Roman forces in Syria, which now included direct administration of Judaea after the removal of Herod Archelaus who had been ruling in Judaea since the death of his father, Herod the Great.

Now, we can see a problem with Luke's citing the census taken under the governorship of Quirinius as the reason for Joseph taking the very pregnant Mary to Bethlehem. Matthew tells us that after King Herod's death the Holy Family returned from Egypt, where they had fled out of fear of Herod, and settled in Nazareth while Herod Archelaus was still ethnarch. This means that Jesus had to have been born before Quirinius took over from Herod Archelaus. Recall also that Luke himself tells us that John the Baptist was conceived during the reign of King Herod (*Lk* 1:5) and that the Incarnation, the visit of Angel Gabriel to Mary, occurred six months later (*Lk* 1:36). Granted, King Herod may have

---

[49] https://en.wikipedia.org/wiki/Coponius

died in the intervening six months, and so Jesus could have been born during the reign of Herod Archelaus. However, this still does not get us anywhere near to Quirinius's census in AD 6-7. So, we are left to consider that Luke may have been wrong on this point. During our period of interest, Augustus Caesar conducted three worldwide censuses. Two of these, the census of 28 BC and the census of 8 BC, occurred during the reign of King Herod. Perhaps Luke confused the Quirinius census with one of these. We may never know. However, as we will see below, there may have been a link in Luke's mind between Quirinius and the birth of Jesus that made him susceptible to making this mistake over the census. In any event, we will call the census an "outlying data point" and promptly ignore it as a possible benchmark by looking at another census.

Syrian governor Sentius Saturninus and the census of 8 BC
The early Church Father Tertullian (born *c.* AD 155, died sometime after AD 220), while writing about Jesus's ancestry in his defence against the Marcionites, mentions, rather matter-of-factly: "But there is historical proof that at this very time a *census* had been taken in Judaea by Sentius Saturninus, which might have satisfied their inquiry respecting the family and descent of Christ."[50]

---

50 Tertullian, *The Five Books against Marcion*, in A. Roberts, J. Donaldson and A.C. Coxe (eds.) and P. Holmes (tr.) Ante-Nicene Fathers – Volume 3: *Latin Christianity: Its Founder, Tertullian* (New York: Christian Literature Publishing Company, 1885) p. 378.

It is possible that Tertullian had information available to him indicating that Jesus was born while Saturninus was governor of Syria, which spanned from 9-7 BC. As mentioned above, Augustus Caesar conducted a worldwide census in 8 BC. Tertullian believes that the records from the census could have settled any questions about Jesus's lineage. Recall that Joseph took Mary to Bethlehem because the census required that people had to return to their ancestral homes to be enrolled in the census. Since Joseph was of the house of David, he took Mary and Jesus (*in utero*) to the city of David, Bethlehem. Both Matthew and Luke tell us that the birth occurred in Bethlehem in order to demonstrate that Jesus was in fact a descendant of King David. Tertullian's remarks provide us with another benchmark, the Saturninus governorship of Syria, 9-7 BC.

Return from Egypt and Jesus in the Temple

Word reached Joseph that "those who sought the child's life are dead" (*Mt* 2:20). Joseph dutifully took his family back to Judaea. "But when he heard that Archelaus reigned over Judaea in place of his father Herod, he was afraid to go there, and being warned in a dream he withdrew to the district of Galilee" (*Mt* 2:22).

When King Herod the Great died in 4 BC, his kingdom was divided amongst his sons. Archelaus took his place in Jerusalem, ruling over Judaea, Samaria and Idumea as ethnarch. Like father, like son: Archelaus's rule was harsh and tyrannical. Not surprisingly, Joseph was reluctant to return to Judaea but instead went to Nazareth in Galilee, which lay outside Archelaus's domain.

Constant complaints about Archelaus's illegitimacy as ruler over the Jews and tyrannical nature moved Emperor Augustus to order Archelaus back to Rome in AD 6. After a trial, Archelaus was relieved of his office and sent into exile in Gaul. Judaea was then annexed to the province of Syria, with Sulpicius Quirinius as legate or – as Luke refers to him – governor.

The sad story of Herod Archelaus does afford us another benchmark. From Matthew 2:22 we know that Jesus was alive during Archelaus's reign. The questions now become these: when did the Holy Family return from Egypt, and how old was Jesus when they came back?

Luke gives a possible clue. In Luke 2:41 we are told that, at the age of twelve, Jesus, Mary and Joseph went to the Temple in Jerusalem. This means that Jesus was "in country" by the age of twelve. If we imagine that the Holy Family with a twelve-year-old Jesus returned to Judaea on the day Archelaus assumed power in 4 BC and went immediately to the Temple, then the earliest date for Jesus's birth would be in the year 16 BC. If the twelve-year-old Jesus went to the temple the year Archelaus was deposed, then his birth would be around 7 BC. If the holy family only spent a year in Egypt and returned on the day Herod died and Archelaus became ruler, then he might have been born as late as 5 BC. Thus, by combining the Holy Family's flight into Egypt with

Luke's report of Jesus being back in Judaea and frequenting the Temple at the age of twelve, we now have a benchmark spanning from 16 BC to 5 BC for the date of Jesus's birth.

Slaughter of the innocents

Our last benchmark to consider has to do with the infamous slaughter of the innocents, as reported in Matthew's Gospel. Matthew reports that the Magi were warned in a dream not to return to Herod with news of the child, and so they returned to their country by an alternative route (*Mt* 2:12). Similarly, Joseph too was warned in a dream that Herod meant to kill the child. His dream warned that, for their safety, Joseph must flee with Jesus and his mother to Egypt (*Mt* 2:13).

As time passed without the Magi returning with news about the child born under a regal star, Herod realised that something was up. In his paranoia, he perhaps thought that the Magi were part of a conspiracy to usurp him of his throne in favour of this newborn king of the Jews, as we read in Matthew 2:16-18:

> Then Herod, when he saw that he had been tricked by the Wise Men, was in a furious rage, and he sent and killed all the male children in Bethlehem and in all that region who were two years old or under, according to the time which he had ascertained from the Wise Men. Then was fulfilled what was spoken by the prophet Jeremiah:

> A voice was heard in Ramah,
> wailing and loud lamentation,
> Rachel weeping for her children;
> she refused to be consoled,
> because they were no more."

As brutal as the slaughter of every male child below the age of two within Bethlehem and the surrounding district is, it is not unimaginable that Herod the Great was capable of giving such an order and making sure it was carried out. As we have mentioned earlier, Herod held onto power partly by being murderously brutal. Whether family or friend, you were not long for this world if Herod perceived you as a threat. Fearing that his own sons, Alexander and Aristobulus, were plotting against him, Herod had them both killed in 7 BC. Then, five days before his own death, Herod had another of his sons executed for attempting to poison him.

Matthew's Gospel is the only extant source for the slaughter of the innocents. Some have suggested that the story is a myth, that it never happened. True, the incident is not mentioned by the first-century AD historian Josephus – our primary, if not only, source of information for this period – but we should not imagine that Josephus chronicled every atrocity committed during Herod's reign of over thirty years. Josephus does tell us that Herod was responsible for killing his family members, so killing children in a small town does not seem out of character.

Matthew's account does, however, have a confusing aspect. He tells us how, based on the time ascertained from the Magi, Herod ordered all children two years old and younger to be killed, ostensibly to make sure that his rival was liquidated. Dr Molnar points that this implies a misunderstanding of astrological principles on the part of Herod or by Matthew. Dr Molnar points out that the threat came only from those born on the auspicious day, or group of days. The inference is that the Magi told Herod that the star appeared two years earlier. Herod then had every male child in Bethlehem and its surrounding area born any time after the star's appearance killed. This apparent misunderstanding of the principles of astrology, Dr Molnar contends, along with the lack of corroborating references to the event from other historical sources, suggests to him that the story is a myth – either one that Matthew created or something that he had simply heard and was passing along.

I would argue that we can trust Matthew in his reporting the slaughter of the innocents as much as we can trust him on any other event he reports in his Gospel. As a matter of fact, some have suggested that the whole star story is itself a creation by Matthew to try and bolster the claim that Jesus was born King of the Jews. It may be argued that both events are unhistorical, that they are fabrications. So, we must ask ourselves, can we point to anything in Matthew's writings that can be identified, with certainty, as a fabrication? I am not aware of anything. As already mentioned, theologian and New Testament scholar Klaus Berger insists, "Even when there is only a single attestation… one must suppose, until the contrary is proven, that the evangelists did not intend to deceive their readers, but rather to inform them concerning historical events… [T]o contest the historicity of the account on mere suspicion exceeds every imaginable competence of historians."[51] Therefore, until we find a trustworthy second source who contradicts Matthew, or a confession by the evangelist saying that he fabricated the star story or the slaughter of the innocents, I will, while maintaining a critical stance, take Matthew at face value.

There is another reason to believe Matthew: public memory. The story of children being murdered in a small town outside Jerusalem may not have "made the cut" in Josephus's history, but it would seem plausible that the event would be remembered for a long time in the region where it happened. Given that Matthew's Gospel may have been written sometime after AD 70, he would risk people contradicting the story if it were not true. This may not be a very strong argument for the authenticity of the story, but it is one more point to consider.

If we accept that the slaughter took place, what are we to make of Herod ordering all males two years old and younger to be killed? This would seem to be inconsistent with what we know about Hellenistic astrology, which would hold that

---

[51] As quoted in Pope Benedict XVI, *Jesus of Nazareth: The Infancy Narratives* (New York: Image, 2012) p. 119.

the only threat to Herod would come from among those born *at the time of* the portent. So, if the Magi told Herod it had been two years since the Star's appearance, then why not kill only the children that were now two years old?

Well, perhaps that is exactly what Herod did.

Herod was a Jew, but he was also a cosmopolitan who understood the Hellenistic culture of the greater world. It is likely that he understood the fundamentals of Hellenistic astrology enough to know that the date of the regal portent's appearance also marked the birth date of his putative rival. Please note that when in Matthew 2:7 we are told how Herod summoned the Magi to meet with him in order to learn the time of the star's appearance, Matthew makes it a point to add that Herod met with the Magi secretly. Why, since it appears that the first meeting was in public, would the evangelist stress the point that the second meeting was held in secret? Perhaps it was because Herod did not want it generally known that he was taking all this astrology stuff seriously.

As already noted, Herod the Great appears to have been an admirer of Hellenistic culture. He gave lavish gifts to the city of Athens and supported the Olympic Games. Josephus tells us how Herod "appointed that solemn games to be celebrated every fifth year, in honour of Caesar, and built a theatre in Jerusalem as also a very great amphitheatre in the plain."[52] Josephus goes on to say that though the theatre and

---

[52] Josephus, *A.J.* 15.268 (Whiston).

amphitheatre were costly works, they were in opposition to Jewish custom. Furthermore, Herod invited competitors from "every nation", that is, the Gentiles, to compete in these games. Having knowledge of and perhaps even belief in Hellenistic astrology would be consistent with Herod's demonstrated admiration and appreciation for Hellenistic culture in general.

Given that Herod understood the fundamentals, and since he ordered the murder of all boys two years old and younger, we should consider that the Magi told him that the time of the star's appearance was one year prior to their secret second meeting. At a time when people did not have birth certificates, how does someone determine if a child is one year old rather than six months old or a year and a half old? It would be difficult to do so with any certainty. Therefore, in order to make sure that he eliminated his one-year-old rival, Herod ordered his forces to kill every boy that was one year old, *plus or minus a year*, which can also be rephrased as all boys two years old and younger.

As we have mentioned, we are closely following Dr Molnar's analysis for the range of possible dates for Jesus's birth. Dr Molnar maintains an agnostic view as to the veracity of the story of the slaughter of the innocents but, nevertheless, includes it as a benchmark in his determination of the birth-date range. He takes the two years and younger to imply that the Magi told Herod that the star appeared two years earlier, which indicates that Herod was alive at least

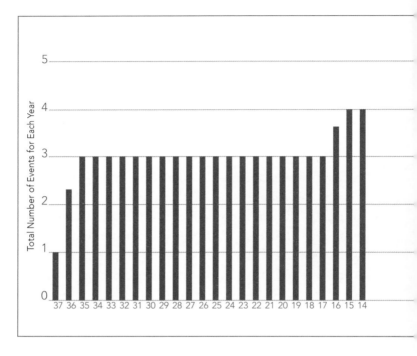

two years after Jesus's birth. For our analysis, we will take the story to indicate that Herod was alive for at least one year after the birth, giving us a benchmark of 36-5 BC.

Summing up

Now that we have a series of benchmarks derived from what we think we know about Jesus's birth, we can look and see where they may overlap. The year or years with the most overlapping benchmarks we can consider the most likely candidates for the year of the Nativity.

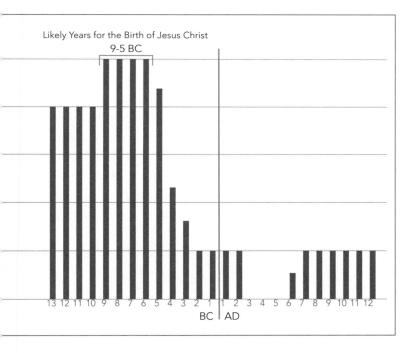

Likely Years for the Birth of Jesus Christ

9-5 BC

13 12 11 10 9 8 7 6 5 4 3 2 1 1 2 3 4 5 6 7 8 9 10 11 12

BC | AD

A plot or graph called a histogram can be useful in helping visualise these overlaps. A histogram is made by first dividing a range into a series of bins or buckets. For our case, we will divide the horizontal axis into fifty-one bins each representing a year. These bins will span a range of years from 38 BC to AD 13. Next, we count how many benchmarks, if any, fall into that bin or year. The value we get for the frequency of events in that year determines the height of the bin.

So, after adding up all the data, we can be reasonably confident that Jesus was born sometime between the years 9-5 BC. As I have been careful to note, we have been following an analysis laid out by Dr Molnar in his study. Reassuringly, Dr Molnar derived a similar range of 8-4 BC. A one-year offset is fully understandable given that most of the historical dates we used for benchmarks we have rounded off to a year. Only for benchmarks that are tied to the death of King Herod, thought to be April of the year 4 BC, have we tried to refine the range to include partial years. These partial years express themselves in our plot wherever a bin does not contain an integer value.

Please note that, off to the far right of the histogram, that lonely looking group of lines is the reign of our friend Quirinius. As mentioned earlier, we will consider his mention in the Gospel of Luke to be an "outlying data point". It is a bit of a joke amongst astronomers that outlying data points are necessary if you want people to take your plots seriously. They are sort of the exception that proves the rule.

We now know where to look for the regal portent – the zodiac sign Aries, the Ram – and we now have a good idea of the period we should be looking at, namely 9-5 BC. However, what exactly are we looking for? But first…

A warning from the Magisterium

All forms of *divination* are to be rejected: recourse to Satan or demons, conjuring up the dead or other practices falsely supposed to "unveil" the future. Consulting horoscopes, astrology, palm reading, interpretation of omens and lots, the phenomena of clairvoyance, and recourse to mediums all conceal a desire for power over time, history, and, in the last analysis, other human beings, as well as a wish to conciliate hidden powers. They contradict the honour, respect, and loving fear that we owe to God alone.[53]

I want to be very clear that, in speaking of astrology, particularly the first-century Hellenistic form of astrology that was in vogue at the time of Christ's birth, I am in no way condoning the practice of astrology. As the above warning states, and it comes straight out of the *Catechism of the Catholic Church*, all forms of divination are to be rejected: primarily because such practices undermine the Christian belief that we are free moral agents who possess free will and are not merely victims to the will of an all-powerful and capricious fate.

However, in order to understand the significance of the star, we must put ourselves in the sandals of the first-century peoples of the Near East who did (with the notable exception of the Jews) believe in fate and the ability of astrologers to predict the future.

The reader might ask, why would God use astrology to further his goals if it is to be rejected by his children? Well,

---

[53] *Catechism of the Catholic Church*, 2nd ed. (United States Catholic Conference, 2000) n. 2116, p. 513.

because the old truism that says that "God can always bring good out of bad" is, indeed, true.

In his book, *The Problem of Pain*, C.S. Lewis observes:

> A merciful man aims at his neighbour's good and so does "God's Will", consciously co-operating with "the simple good". A cruel man oppresses his neighbour, and so does simple evil. But in doing such evil, he is used by God, without his own knowledge or consent, to produce the complex good – so that the first man serves God as a son, and the second as a tool.

The reader may object to my implication that the Magi were somehow cruel or evil. Am I not being a little rough on them? They were just benighted pagans, after all, who probably did not know better. How can we say that their belief in astrology made them cruel or oppressing?

Well, to quote again from the *Catechism of the Catholic Church*: "Consulting horoscopes, astrology, palm reading, interpretation of omens and lots, the phenomena of clairvoyance, and recourse to mediums all conceal a desire for power over time, history, and, in the last analysis, other human beings..."[54] I am not saying that the Magi could not have been "nice fellows"; however, they were trying to gain an advantage over people. A newborn king implied that there might be changes in the regime. Perhaps new job opportunities would open up. It was, after all, the rich and

---

[54] *Ibid.*, p. 513.

powerful who would keep astrologers "on staff" in order to "have power over… other human beings". So, perhaps the Magi, like all of us for that matter, were cruel in a subtle and nuanced way.

Setting aside whether or not the Magi were "nice fellows", it remains that God can bring a complex good out of that which is objectively evil. To paraphrase C.S. Lewis, the Magi may not yet have been God's sons, but they were his tools.

There is another temptation to avoid as well. We might think that because the Magi read a message in the stars that we, as Christians, know to be true, we have to allow for the possibility that astrology 'works'. However, recall how different the Hellenic understanding of the flow of time and the existence of freedom truly were from Judaeo-Christian thought on the subject: their cosmology made it easy to believe that the stars caused whatever happened on earth, and because the motion of the stars was measurable and predictable, this only added to the suspicion that free will was at best an illusion. Conversely, from a Christian perspective, what occurred when the Magi saw, interpreted and followed the Star of Bethlehem was, if not simply a coincidence, a sort of anti-astrology. As Pope Benedict XVI so insightfully noted, "… it is not the star that determines the child's destiny, it is the child that directs the star."[55] This particular child, unlike any other ever born, was the Word of God –

[55] Pope Benedict XVI (2012) *Jesus of Nazareth: The Infancy Narratives* (Image, Random House, New York, p. 101) Liberia Editrice Vaticana

the *Logos*, the very agent of creation – which means that the arrangement of the stars didn't *cause* his birth in Bethlehem, but rather, by whatever means he created the universe, he quite literally put the stars in the heavens and governed their motion by laws of nature that determined that they would reach those precise positions in the sky at the point in time when the Word became flesh. The stars didn't 'move' the child; *the child moved the stars*. The reader can believe that this was a coincidence, or that it was some sort of divine cosmic 'easter egg'[56] worked into the design of Creation in the very beginning. It doesn't really matter either way so far as the efficacy of astrology is concerned. There is no logical reason to jump from one correct inference according to one arcane system – which prediction, to be clear, just happened to be about the person who designed the thing the system measures – to the belief that the stars can predict anything about the lives of ordinary people like us. In other words, the system was rigged.

So, you have been warned: stay away from astrology!

---

[56] 'Easter eggs' are little jokes or messages or extra hidden features left in a digital game or application by its designers. They're not a part of the program's documentation or core processing, so they aren't intrinsic to how the game or application functions. Easter eggs are usually weird, wonderful, or whimsical and serve mainly to delight and amuse 'serious players and users.'

What Makes a Portent Regal?

Now, to get back to our search for a portent of a regal birth.

Dr Molnar tells us that the "principles of Greek astrology concerning regal births are not intuitively obvious, and even a quick perusal of the subject reveals that Greek astrology is complex and arcane". So we are only going to look at the principal indicators and, specifically, how they would have to be arranged to point unambiguously at the birth of a king in Judaea.

There are many aspects to a regal birth, but we will, mercifully, limit ourselves to examining five major ones:

1. Exaltations
2. Rulers of the trines
3. Attendant planets
4. Masculine signs
5. Conjunctions and occultations

Keep in mind that, though each of these aspects is necessary for a regal portent, it is the cumulative effect of all these portents in one natal horoscope that makes for an iron-clad portent of a regal birth.

Astrologers would use these various aspects of a regal birth to analyse star charts, what we call horoscopes. These charts would depict the position of the zodiacal signs and the relative placement of the Sun, Moon, planets and bright stars for the date, time of day and location of the birth in question. Each planet was thought to have properties that could help manifest the destiny Fate had in store for the individual, from their birth to their death.

We should also note that because of advances made in mathematical astronomy, astrologers could calculate the positions of the planets and were no longer restricted to only what they could observe. In particular, the positions of planets made invisible because of daylight could now accurately be known and their position relative to the zodiac ascertained. Perhaps even more importantly, astrologers at the dawn of the first millennium could draw up a chart or horoscope for anyone born anywhere and at any time – past, present or future.

Determining the position of the Sun, Moon, planets and bright stars relative to the zodiacal signs on the day of the birth was one job of the natal star chart. Its second job was to determine the position of the planets at the time of the birth relative to the local horizon. The entire sky, both above and below the horizon, was considered in forecasting a horoscope. Of particular interest was the portion of the zodiac and any planets that were ascending – that is, rising in the east.

Star charts or horoscopes in ancient times were drawn as a box divided into twelve sections, each representing one of the twelve signs of the zodiac. A notable example is the natal astrological chart for Emperor Hadrian drawn in the second century AD to help him solidify his claim to the imperial throne.

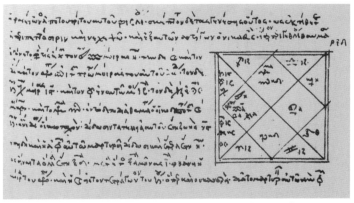

An example of how astrological charts were drawn in the second century AD. This one is Emperor Hadrian's natal horoscope.

On the next page you will find the same chart drawn in modern form: the horizontal line is the horizon as we face south. East is to the left and west is to the right. The great circle is the zodiac, divided into its twelve signs and oriented, as the signs would be observed, relative to the horizon, at the particular place and time of the birth. The

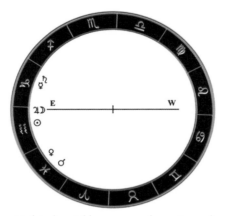

Emperor Hadrian's natal horoscope redrawn in modern form.

symbols inside the ring represent the Sun (☉), Moon (☽) and planets. Astrologers tend to call all these objects planets or stars, the Sun and Moon included.

Please note: this symbol, ♈, represents Aries, the ram. For the case of Jesus, any regal portents would involve the zodiacal sign, Aries.

Exaltations

The first of our regal portent indicators is exaltations. An exaltation is the zodiacal sign in which a planet takes on omnipotent power and, thus, is exalted. This diagram shows the Sun and each of the other planets at their positions of exaltation. Note: the Sun is exalted when it is in Aries (♈).

Rulers of the trines

Next come the rulers of the trines. Trines are formed by any three zodiacal signs that are separated from each other by

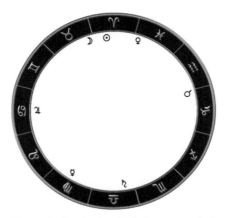

Planets in the sign in which they are exalted.

120 degrees. The twelve signs of the zodiac can be arranged into four distinct trines. The trine shown here is the one that contains Aries (♈), along with its trine partners Leo (♌) and Sagittarius (♐).

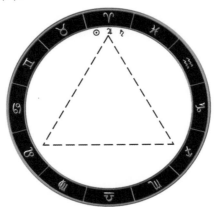

Trine I (Aries, Sagittarius, Leo) and its Rulers: Sun, Jupiter, Saturn

Each of the four trines was given three controlling planets called rulers. The presence of rulers in their respective trines was a powerful condition for a regal portent. In this trine, the rulers are the Sun, Jupiter (♃) and Saturn (♄). These planets could be arranged any number of ways, but this diagram shows the optimum one. This alignment would happen about once every sixty years and so is a "once in a lifetime event". The Magi would have been full of admiration at how the conditions for the rulers of the trine had been superbly met.

Attendant planets and masculine signs
Attendant planets are next on our list. These planets, by their proximity to the Sun or Moon, ward off the deleterious effects of badly placed planets. Jupiter and Saturn are the Sun's attendants and should precede the Sun – that is, they should rise before the Sun in the morning. Venus (♀) and Mars (♂) attend the Moon and should follow the Moon – that is, they should rise after the Moon.

Another requirement for a regal portent is that the Sun or Moon should be in one of the zodiac's masculine signs. The masculine signs are Gemini (♊), Leo (♌), Libra (♎), Sagittarius (♐), Aquarius (♒) and, as the reader may have already suspected, Aries the Ram (♈).

Summary: regal portent for Judaea
Let us pause for a moment and summarise the aspects that must be present for a portent of a regal birth as it applies

specifically to Judaea. First and foremost, our ancient authorities assure us that Judaea fell under the zodiacal sign Aries, the ram. Any regal portent would be expected to occur in Aries. Since Aries is one of the masculine signs, then we have our first aspect of a regal birth satisfied. Now, if the Sun should be in Aries, it would be exalted, taking on omnipotent power, which would fulfil another requirement. If Jupiter and Saturn were in attendance with the Sun in Aries, rising before the Sun in the morning, then the Sun would be protected from any "badly placed" planets. Furthermore, all three rulers of the trine, the Sun, Jupiter and Saturn, would be together in their most auspicious configuration and so another aspect of a regal portent would automatically be satisfied.

By now your head must be swimming, but remember that we are only considering some of the more important aspects for a regal birth. It is all much more complicated than this.

Now, to us moderns, astrological signs, exalted planets and rulers of the trines all seem pretty arbitrary and sound pretty silly. As Dr Molnar says, "The principles of Greek astrology concerning regal births are not intuitively obvious, and even a quick perusal of the subject reveals that Greek astrology is complex and arcane."[57] The arcane and opaque nature of first-century Hellenistic astrology may be a reason

---

[57] Molnar, Michael R., *The Star of Bethlehem: The Legacy of the Magi* (New Brunswick, NJ: Rutgers University Press, 1999) p. 64.

why people have looked for other, simpler explanations for the Star of Bethlehem. However, we must keep in mind that the people of the Near East at the beginning of the first millennium (with the notable exception of the Jews) took all this stuff very seriously; especially, those at the highest levels of society as exemplified by Emperor Hadrian having his natal horoscope cast. So, let us persevere.

Dr Molnar contends that, for first-century Magi, this portentous configuration, the Sun, Jupiter, and Saturn all in Aries would, on its own, serve as an unambiguous indication of a regal birth in Judaea. But the evangelist Matthew speaks of a single star and not a group of stars. So, is there a configuration of regal principles that come together in such a way as to have one star play the dominant role?[58]

Conjunctions and occultations

The last item on our list of regal portents to be considered is conjunctions and occultations. Conjunctions occur when celestial bodies come in close proximity to one another. Occultations are when one body passes in front of another. Solar and lunar occultations, referred to as eclipses, have often been seen as powerful omens. As we have already noted, the ancient historian Herodotus claims that when an eclipse occurred during a battle between the Medes and the Lydians, who were starting their sixth consecutive year of conflict, they quickly stopped fighting and came to peace

---

[58] *Ibid.*, p. 77.

terms. Herodotus wrote, "The Lydians and Medes, seeing night succeeding in the place of day, desisted from fighting, and both showed a great anxiety to make peace."[59]

As impressive and awe-inspiring as a total solar eclipse can be, first-century astrologers believed that conjunctions between less visually impressive celestial bodies were also very important. As noted earlier, what made for a spectacular regal portent was the number of regal principles present and conditions met in a horoscope. One regal principle that might not be as visually arresting as an eclipse is the proximity of the Moon with Jupiter.

When Greek astrologers spoke of "stars", this also included the Sun, Moon and planets. The powers ascribed to a star paralleled its namesake's role in Greek mythology. As such, it is not surprising that Jupiter [Zeus] is seen as an indicator of fame, nobility and kingship. As powerful a regal birth indicator as Jupiter was on its own, the influence of the planet was amplified by its proximity to the Moon. Should the Moon actually occult Jupiter, well, that would hit the ball right out of the Colosseum. So, if the Moon occulted Jupiter as Jupiter attended the Sun in Aries, then this could have signalled to the Magi that a king had been born in Judaea, and Jupiter would have played the critical role.

Knowing that it was probable that Jesus was born sometime between 8 BC and 4 BC (9 BC-5 BC by our

---

[59] Cary, Henry, *The Histories of Herodotus*, (New York: Appleton & Co., 1899) pp. 28-29.

analysis), Dr Molnar cranked up his computers to see if, indeed, the Moon had occulted Jupiter in the sign of Aries during this time period. Amazingly, not one but two lunar occultations of Jupiter occurred in Aries during this period, and both occurred in the year 6 BC, during the reign of King Herod, within our estimated range of years for Jesus's birth.

Just before sunset in Judaea on 20th March 6 BC and then again, a month later, on 17th April 6 BC, the Moon occulted Jupiter while in Aries. But which, if either, of these events would the Magi have considered *the* star?

It Is All Greek to Me

Let me first begin by admitting that I know little to nothing about New Testament Greek. Pretending that I have any expertise in Koine Greek, also known as Hellenistic Greek and Biblical Greek, well, that would be going too far. That said, I will be relying heavily on the work of Dr Molnar.

The following is a word for word translation under the original Greek text of Matthew 2:9.[60]

καὶ ἰδοὺ ὁ   ἀστήρ, ὃν   εἶδον   ἐν τῇ ἀνατολῇ,
and look the star   which they saw   in the east

προῆγεν αὐτούς, ἕως ἐλθὼν   ἐστάθη ἐπάνω
led before them   until having come it stood upon

οὗ   ἦν τὸ παιδίον
where was the small child

We have already noted that this star is peculiar. The Magi observed it rising in the east, but it then leads them as they travel west, a seeming contradiction. Then, amazingly, the star stands over the child, a patently unusual celestial

---

60 Nestle, E. and McReynolds, P.R., *Nestle-Aland 26th Edition Greek New Testament with McReynolds English Interlinear* (Logos Research Systems, Inc., 1997) (*Mt* 2:9).

event. All this confusion, Dr Molnar contends, arises from thinking like a modern and not as a first-century magus.

The phrase "in the east" is a literal interpretation of the Greek "ἐν τῇ ἀνατολῇ", which actually means "at the rising". Specifically, for Greek astrologers, it meant a "heliacal" rising. A heliacal rising is when a celestial object rises just prior to sunrise. If this sounds familiar, we already mentioned how the Egyptians watched for the heliacal rising of the star Sirius to know when the Nile would soon begin to flood. It is not difficult to work out that heliacal risings are very important to astrologers.

So, when, if at all, did Jupiter have a heliacal rising in Aries during our time frame? Only once, Dr Molnar tells us; on the morning of 17th April 6 BC there was a heliacal rising of Jupiter. Recall that this is also the exact date of the second lunar occultation of Jupiter.

The Magi, upon seeing a heliacal rising of Jupiter in Aries, on the same day as a lunar occultation of Jupiter in Aries, while the Sun was exalted – in Aries – with attendant planets all in place to ward off any possible deleterious effects of badly placed planets, would have interpreted this as a stupendous portent pointing to a regal birth in Judaea.

But what about the star that "went before" and that was "standing over"?

To answer that question, let us read Matthew as a magus would. According to Ivor Bulmer-Thomas in an article about the Star of Bethlehem appearing in a 1992 issue of the *Quarterly Journal of the Royal Astronomical Society* (here quoted from Dr Molnar's book), the word Matthew used that is translated into English as "went before" is the word προῆγεν (*proegen*). This word, however, is related to the astrological term προηγήσεις (*proegeseis*), which indeed means "to go before", but more precisely "to go in the same direction as the sky moves".[61]

Recall from our astronomy discussion how, from night to night, the stars appear to drift from east to west, while the planets normally drift in the opposite direction, from west to east as they orbit the Sun. Occasionally, a planet may stop its eastward drift, pause, then drift westward "in the same direction the sky moves", pause again and then return to its original easterly course. This stopping, reversing direction, stopping again and then returning to a normal eastward drift is most easily observed for Mars, Jupiter and Saturn.

When a planet changes direction and backtracks towards the west, modern astronomers call this "retrograde motion". First-century magi did not use the term "retrograde", but instead they called this προηγήσεις (*proegeseis*). But Matthew used the related though non-technical term προῆγεν (*proegen*). This word does indeed mean "went before", but in using it, Matthew obscured the astrological meaning, namely "to go in the same direction as the sky moves".

---

61 Molnar, Michael R., *The Star of Bethlehem: The Legacy of the Magi* (New Brunswick, NJ: Rutgers University Press, 1999) p. 90.

Lastly, the phrase "stood over" has been another cause for misunderstanding. Bulmer-Thomas explains how the Greek word ἐπάνω (*epano*) does, in some circumstances, mean "over". However, our guides to first-century Hellenistic astrology, such as Claudius Ptolemy of Alexandria, use this word to mean "above" as in "above in the sky". If we interpret the word astrologically, then it means that the planet became stationary "in the sky", which would, coincidently, be over the child.[62]

All in all, verse 2:9 from Matthew's Gospel has strong astrological implications. Unfortunately, the astrological nature of what was being reported became obscured when the evangelist chose to use laymen's terms rather than the technical vocabulary of astrologers. Reinterpreting Matthew 2:9 astrologically is consistent with the motions of the planet Jupiter when it rose in the east on the morning of 17th April 6 BC and in the months following.[63]

On this modern star chart, with the boundaries of Aries (♈) marked out, there are symbols showing where each planet was on 17th April. By convention, east is to the left on star charts. We astronomers do that to confuse the uninitiated. Note how the Moon (☽) and Jupiter (♃) are in occultation in the middle of the sign. Mars (♂) is to the left in the neighbouring sign of Taurus. The Sun (☉), Jupiter (♃),

62 Ibid., p. 92.

63 Ibid., p. 93f.

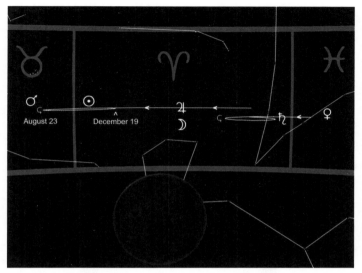

Jupiter's retrograde motion in the year 6 BC traced over a star chart of the constellation Aries. Planets are shown at their positions for 17th April 6 BC

the Moon (☽) and Saturn (♄) are all in Aries. Venus (♀) is to the right in the astrological sign of Pisces.

Most of the planets were too close to the Sun to be visible on 17th April; however, after a few days, as the Sun drifted into Taurus, Jupiter became visible as it rose in the east as a morning star.

Jupiter drifted eastward, leaving Aries and entering Taurus around 20th June. It then reached its first station on 23rd August and stood motionless in the sky for about a week as the earth caught up with it. Jupiter then reversed direction and proceeded westward, *in the direction that the sky moves*, which in the parlance of the Magi is προηγήσεις (*proegeseis*), but what Matthew called προῆγεν (*proegen*), "went before". Jupiter stood still a second time, reaching its second station on 19th December in Aries – the astrological sign representing the place of the newborn king's birth. Matthew describes this as the star that "came to rest over the place where the child was" (*Mt* 2:9b).

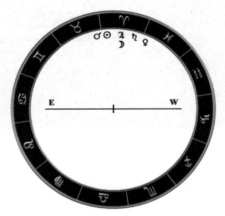

Occultation of Jupiter by the Moon in Aries, 17 April 6 BC

Dr Molnar points out that the Magi would have celebrated this second stationing of Jupiter in Aries because the principle of the trine rulers had once again been fulfilled: the Sun in Sagittarius, with Jupiter and Saturn in Aries. Nevertheless, this is only a faint echo of the far more important conditions present on 17th April.[64]

Computer simulation of the sky as it appeared at sunrise on 17th April 6 BC, facing south, near Jerusalem.

We can now use Dr Molnar's idiomatic paraphrasing of Matthew 2:9 from an astrological point of view and so recap all that we have discussed thus far:

And behold, the planet Jupiter, which they had seen at its heliacal rising, went retrograde and became stationary above in the sky (which disclosed) where the child was.[65]

---

[65] Ibid., p. 96.

Using computer software that can simulate the appearance of the sky for a given time and place we have created this image of how the early morning sky appeared near Jerusalem on 17th April 6 BC. This is the configuration of planets that made the Magi so excited and Herod so afraid.

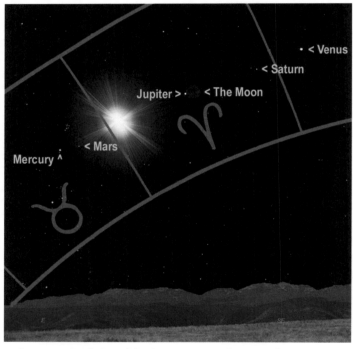

Computer simulation of the sky as it appeared at around 8:10 a.m. on 17th April 6 BC, facing south, near Jerusalem. Atmosphere removed and the band of the zodiac outlined.

In software, we can make it easier to see the stars by removing the atmosphere with its clouds and scattered sunlight. We can also label the Moon and planets and outline the zodiac and signs.

The reader may notice that the marker for the Moon seems to be pointing at nothing. The Moon is there, though, just to the southeast (right) of Jupiter. Because the Moon is in close proximity in the sky to the Sun, it is a *new* Moon. By running our simulation forward to 8:10 a.m., local time, we can see the Moon begin to occult Jupiter in Aries.

This could well be the celestial event that Matthew attempted to report on in his Gospel.

What Did the Magi See?

The Star of Bethlehem has been thought to be many things. Some said it was a comet, others a supernova or some such spectacular astronomical event. Still others claimed it was a mere fable, a story of a portent made up by Matthew to give Jesus's birth some gravitas. Still others imagined the star as an angel or fairy flittering about the sky leading the Wise Men to the scene of the nativity.

I believe the Magi themselves are the key to understanding the Star of Bethlehem. The Magi were Wise Men. They explained natural phenomena and divined from omens what Fate had in store for their clients. They were astrologers. Therefore, the Star of Bethlehem is a complex of astronomical phenomena that came together to announce a regal birth of great significance in Judaea.

We have seen how Matthew's perplexing and enigmatic passage about a star dancing around the sky can make perfect sense. It can make sense, that is, if we remember that Matthew is a Jew who knew little or nothing about astrology. Yet, he tries to convey to his Jewish readers what he understands to be the important points of an astrological apparition accompanying Jesus's birth.

First-century astrology was not only subtle, but it was also arcane and complex. It had its own vocabulary of technical terms. So, then, we should not be surprised that Matthew's layman's description of the Star of Bethlehem came out a little confused. Nevertheless, if we keep in mind that Matthew is attempting to describe the highlights of an astrological event, then we can try and recover what he meant to say. Viewed from the position of knowing a bit about Hellenistic astrology, Matthew's text can make sense – that is, make sense to a first-century astrologer.

But this raises a serious question: why does Matthew, a Jew writing about the birth of the long-awaited Jewish Messiah for a readership who are presumably mostly Jews, write anything at all about astrological portents associated with the birth? Astrology, as we have repeated many times, was anathema to Jews of the first century, as it is to Christians in the twenty-first century. Of course, Gentile rulers, in their ignorance of the God of Israel, the God of History, would enthusiastically publicise any astrological evidence showing that Fate ordained their right to rule – but surely never the Jewish Messiah? Surely it is unseemly to even consider God allowing astrological speculations to be associated with the Incarnation?

Well, there may have been a good reason why Matthew *had* to include the star in his account of the Messiah's birth.

This reason might also shed some light on why Luke might have got the census story wrong. But before we get to that, let us first consider why there is a star depicted on the coins.

Dr Molnar has identified a truly unique and impressive regal portent that would have sent the Magi rushing off to Judaea in search of a new king. It occurs within the time period we would expect for Jesus's birth. We have untangled Matthew's valiant attempt to write down what he understood to be the astrologically significant aspects of that regal portent. We can understand that, since Matthew was a Jew, who presumably knew little if anything about astrology, the words he chose only served to obscure the original astrological meaning.

We also now understand why the Magi saw the star, and yet King Herod and all of Jerusalem did not. The early morning occultation of Jupiter by the Moon in Aries was not visible. It was lost in the Sun's glare. Only because the Magi possessed the ability to calculate celestial positions, could they have known that the Moon had occulted Jupiter during the daytime.

Furthermore, as we have said over and over, almost everyone in the Roman world of the first century, especially in the east, followed astrology – that is, with the notable exception of the Jews. Therefore, since the Jews did not follow astrology, they would not give any significance to, nor would they care about, a lunar occultation of Jupiter in Aries on 17th April 6 BC. Not, that is, until the Magi came calling.

Herod's reaction to the Magi's message indicates that, even though a Jew, he believed their conclusion that the star signified that a king had been born in Judaea. The detail, "Then Herod summoned the Wise Men *secretly* and ascertained from them what time the star appeared" (*Mt* 2:7a) allows us to infer that Herod truly believed in the efficacy of astrological predictions concerning future events. He also understood that the Magi's ability to read portents could provide him with useful political intelligence in order to maintain his control on power. But good old King Herod was not the only person who feared the birth of a Jewish Messiah who might one day lead his people in revolt against their rulers. Herod's puppet masters in Rome were most certainly aware of the stunning regal portent that occurred in Aries.

The Romans knew their Jewish subjects to be a restive people. It had just been a little over a hundred and fifty years since the Maccabean Revolt. Indeed, the Romans may have played a role in tipping the balance in favour of the Jews overthrowing their Syrian rulers. If Rome helped the Jews throw off Syrian rule, then someone else might aid them in throwing off Roman rule.

The Romans were also well aware of a conviction held by their Jewish subjects that a Messiah, a king, was to appear who would lead them to independence from foreign rule. For eleven years, now, Jerusalem had been abuzz with rumours that a sign had appeared in the heavens signalling the birth of this long-expected king and deliverer. With the passage of years, this putative Messiah was getting older and would soon come into his manhood. Persistent rumours of a regal birth coupled with a restive people with a history of rebellion – this definitely required that some action be taken by local authorities to mollify the situation.

Getting now to the heart of the question why there is a star depicted on the coins, recall how we discussed the way that "Roman provincial coinage served as a primary medium for disseminating propaganda supporting the goals of Rome."[66] Consider, too, that in AD 6, when, by our calculations, Jesus would have been around eleven years old, Rome deposed Herod Archelaus while simultaneously placing Judaea, Samaria and Idumea under the direct administrative control of Quirinius, governor of Syria, and that, around the same time, new coins began to be struck in Antioch and the other cities in the province. In the light of all that we have learnt, it now seems more likely that these new coins, depicting a leaping ram looking over its shoulder at a star, were intended to subvert the Star of Bethlehem and twist its meaning for the benefit of the emperor. The intended message may have been something like this:

> Hear ye, hear ye! You know of a regal portent in Aries signifying that a king has been given unto you. Hear this, then! The new king's name is Caesar. Behold, I, Caesar

66 Ibid., p. 48.

have deposed the ethnarch, Herod Archelaus, and annexed Judaea to Roman Syria. The Kingdom of Judah has been eradicated from the face of the Earth as proof of my power.

Or some stentorian proclamation along these lines.

Governments have been known to bend the truth to suit their needs. They have even given historical events a new spin, a new narrative to help the common folk deal with the "New Normal". Referring specifically to the use of Imperial Roman coinage to help advance the regime's programmes, classicist and numismatist Michael Grant says, "They [the coins] provide pious hopes, wishful thinking, and downright lies – a common feature of propaganda at many epochs."[67]

So, if this is true and the coins were intended to counter widespread rumours of an astrological portent in Aries signalling the birth of a Jewish Messiah, then we now have tangible proof that Matthew's star is an historical event.

Now back to how these coins may also explain two peculiarities in the Gospels, namely, why does Matthew, a Jew, include a taboo topic such as astrology in his story of the Messiah's birth, and how did Luke mistakenly believe that Jesus was born when Quirinius was governor of Syria?

First, it is very likely that Matthew's Jewish readers had these coins in their purses. If these coins were known to be

---

[67] Grant, Michael, *Roman History from Coins: Some Use of the Imperial Coinage to the Historian* (Cambridge: Cambridge University Press, 2010) p. 69.

an attempt by their Gentile overlords to subvert a widespread belief in a regal portent heralding the Messiah's birth, then Matthew's readership would expect him to address the regal portent in the story. His readership would demand that Matthew provide a rebuttal to the Roman propaganda.

Matthew's goal in the opening chapters of his Gospel is to establish Jesus's credentials as the Messiah. Matthew begins by first laying out Jesus's genealogy (*Mt* 1:1-17). He goes on to show how Jesus's birth fulfils prophecy:

"A virgin shall conceive and bear a son…"(*Mt* 1:22-23)
"The Messiah will be born in Bethlehem, in the land of Judah…" (*Mt* 2:3-6)
"Out of Egypt have I called my son…" (*Mt 2:15*)
"A voice was heard in Ramah, wailing and loud lamentation…" (*Mt* 2:17-18)
"He shall be called a Nazarene…" (*Mt* 2:22-23)

Demonstrating how the birth of Jesus of Nazareth fits in with the prophecies from Hebrew scriptures is to be expected, but Matthew then goes on to show how the God of Abraham even orchestrated the false beliefs of the Gentiles to give witness:

Now when Jesus was born in Bethlehem of Judaea in the days of Herod the king, behold, Wise Men from the East came to Jerusalem, saying, "Where is he who has been born king of the Jews? For we have seen his star in the East and have come to worship him." (*Mt* 2:1-2)

...and behold, the star which they had seen in the East went before them, till it came to rest over the place where the child was. When they saw the star, they rejoiced exceedingly with great joy; and going into the house they saw the child with Mary his mother, and they fell down and worshiped him. Then, opening their treasures, they offered him gifts, gold and frankincense and myrrh. (*Mt* 2:9b-12)

We can imagine how Matthew's Jewish audience might delight in hearing about the Magi, these Wise Men, these Gentiles, bowing down and worshipping the incarnate God of Israel. Matthew's rebuttal of Roman propaganda downplaying the birth of a Jewish Messiah would strengthen his argument that Jesus was in fact the Messiah. I contend, that far from being scandalised, Matthew's audience would be amused by the futility of their rulers railing against the Christ with their silly little coins.

Next, Luke's confusion over who was governor of Syria at the time of Jesus's birth and what census was taking place (if any) might also be linked to the coins. These coins first appear in Antioch during or slightly after Quirinius's governorship. Scholars believe that the evangelist Luke is from the city of Antioch. It is thought that Luke's Gospel is written between AD 80-110. The coins were still being minted in various cities around the province at the time Luke is writing. These coins have been in circulation throughout his lifetime. We can speculate that the association of the coins with the Star of Bethlehem and their link with Quirinius could have led Luke to mistakenly connect the Nativity with Quirinius's governorship.

There is one last historical curiosity that might provide a little more substantiation that the coins may be linked to the Nativity. Antioch uses Aries on their coins for nearly two centuries, well into the mid-third century AD. Reportedly, they died out because of civil wars.[68] The tumult of the third century, referred to as the Military Anarchy or the Imperial Crisis, nearly destroyed the Roman Empire. It seems plausible that in the commotion, minting of provincial coins in Syria may have been affected.

Curiously, though, the mid- to late-third century AD also coincides with the Little Peace of the Church. Eusebius of Caesarea, a fourth-century bishop and a Greek historian of Christianity, tells of a forty-year period in the late-third century without official persecution of the Church. The Peace is called "little" to differentiate it from the Peace of the Church resulting from the Christian conversion of Constantine the Great. Emperor Gallienus, in AD 259, promulgated the first declaration of tolerance with regard to Christians. So, could the relaxation of hostility towards Christianity, during the Little Peace, possibly have something to do with the fading away of the polemical anti-Christian coins? Perhaps.

---

[68] Molnar, Michael R., *The Star of Bethlehem: The Legacy of the Magi* (New Brunswick NJ: Rutgers University Press, 1999) p. 152, n. 34.

As I said in the Introduction, whether or not the coins are truly linked to the star we will never know with metaphysical certitude. C.S. Lewis reminds us that all historical facts are taken on authority. We believe them true because we find that the people who tell us so are themselves trustworthy. We find the preponderance of evidence to be consistent with the historical fact being true.

＊＊ ＊＊ ＊＊

The coins exist. The ram and the star are purposefully placed on the coins to communicate a message. The people who designed the coins believed the general public would readily understand what was being said by these symbols. As we noted earlier, those who will not be God's sons will become his tools. By striking these coins, the enemies of the Christ Child, quite unwittingly, left behind for us tangible, physical, historical evidence that a Star once shone over a little town in Judaea the day the King of the Universe was born.

> And thou, that art still in thy cradle,
> The sun being crown for thy brow,
> Make answer, our flesh, make an answer,
> Say, whence art thou come – who art thou?
> Art thou come back on earth for our teaching
> To train or to warn –?
> Hush – how may we know? – knowing only
> A child is born.

> G.K. Chesterton, The Nativity

Epilogue

Have you ever wanted to experience the world just as Jesus, Mary and Joseph knew it? Have you wanted to look out and see the same things they saw two thousand years ago?

Well, then, this is what I want you to do. On the next clear night, travel away from the glow of city lights to a rural spot with good horizons and look up. What you will see spread above you is the night sky appearing almost exactly as the people of Old and New Testament times saw it.

The shepherds in the fields watching their flocks by night would recognise the constellations that you see and probably call many of them by the same names, albeit in a different language.

Go out tonight and look up, for "the heavens proclaim the glory of God and the firmament shows forth the work of His hands" (*Ps* 19:1).

Image Credits

Cover image: *Star of Bethlehem* by Elihu Vedder (1879), public domain, CC0 1.0 Wikimedia Commons.

Page 6: Sculpture from the facade of Amiens Cathedral © Lawrence Lew O.P.

Page 20: Halley's Comet on the Bayeux Tapestry, public domain, CC0 1.0 Wikimedia Ccommons.

Page 24: Supernova RCW 86, X-ray: NASA/CXC/SAO & ESA; Infared: NASA/JPL-Caltech/B. Williams (NCSU).

Page 87: Bronze coin © Molnar.

Page 107: Histogram © Fr Douglas McGonagle.

Page 115: Hadrian's natal horoscope, used with permission of the American Philosophical Society.

Page 116: Modern version of Hadrian's horoscope © Molnar.

Page 117: Planets shown in their exultations © Molnar.

Page 117: Rule of trines and Judea © Molnar.

Page 127: Retrograde motion of Jupiter © Molnar.

Page 128: Occultation of Jupiter by the Moon in Aries © Molnar.

Page 129: Simulation of sky over Bethlehem © Fr Douglas McGonagle.

Page 130: Zodiac over Bethlehem © Fr Douglas McGonagle.